Australian Sheep Husbandry

A Handbook of the Breeding and Management of Sheep

by Albert Stapleton Armstrong

with an introduction by Jackson Chambers

Self Reliance Books

Get more historic titles on animal and stock breeding, gardening and old fashioned skills by visiting us at:

http://selfreliancebooks.blogspot.com/

Introduction

I am pleased to present yet another practical title on breeding and raising livestock.

The work is in the Public Domain and is re-printed here in accordance with Federal Laws.

As with all reprinted books of this age that are intended to perfectly reproduce the original edition, considerable pains and effort had to be undertaken to correct fading and sometimes outright damage to existing proofs of this title. At times, this task is quite monumental, requiring an almost total "rebuilding" of some pages from digital proofs of multiple copies. Despite this, imperfections still sometimes exist in the final proof and may detract from the visual appearance of the text.

I hope you enjoy reading this book as much as I enjoyed making it available to readers again.

Jackson Chambers

Respectfully Dedicated

TO

ANDREW RUSSELL INGLIS, Esq.

OF

MELBOURNE, VICTORIA

PREFACE.

———◆———

IN offering this work to the Australian public, the authors feel that a few prefatory remarks are necessary.

The object of the book is to supply squatters, station-managers, and selectors with information, in as concise a form as possible, regarding sheep and their diseases, and all branches appertaining to sheep husbandry in Australia, including suggestions as to the best quality of sheep to breed, and as to the general management of stations; complete descriptions of tank and well-sinking, fencing, and dam-making, with their approximate cost, and rules for their measurement; and also methods for the destruction of dingoes, kangaroos, &c.

There are many subjects treated, upon which a variety of opinions exists, and in dealing with these, the authors, while giving due consideration to the opinions of others, have advanced their own without, they hope, any appearance of dogmatism.

PREFACE.

Although they have written from a long and varied experience among sheep stations, the authors have deemed it advisable, in order to render the work as valuable and interesting as possible, to quote freely from other authors. When doing so, however, they have been careful to acknowledge their indebtedness, and now tender their thanks to Messrs. Randall, Youatt, and others. A quantity of information, such as statistics, &c., has been taken from the Encyclopædias, and in these instances no acknowledgment has been made.

No doubt numerous errors of form and construction may be discovered in this book, but the authors offer no apology for these, deeming their work to be more of a practical than a literary character.

MOOTHUMBIL, LACHLAN RIVER,
·October, 1881.

CONTENTS.

CHAPTER I.

CHAPTER II.

CHAPTER III.

CHAPTER IV.

CONTENTS.

CHAPTER V.

CHAPTER VI.

CHAPTER VII.

CONTENTS.

CHAPTER VIII.

CHAPTER IX.

CHAPTER X.

CHAPTER XI.

CHAPTER XII.

CONTENTS.

CHAPTER XIII.

CHAPTER XIV.

CHAPTER XV.

CHAPTER XVI.

CHAPTER XVII.

CONTENTS.

CHAPTER XVIII.

CHAPTER XIX.

CHAPTER XX.

CHAPTER XXI.

CHAPTER XXII.

CONTENTS.

CHAPTER XXIII.

CHAPTER XXIV.

CHAPTER XXV.

CHAPTER XXVI.

CHAPTER XXVII.

CHAPTER XXVIII.

CHAPTER XXIX.

AUSTRALIAN SHEEP HUSBANDRY.

CHAPTER I.

AUSTRALIA.

AUSTRALIA is essentially a wool-producing country. If we compare the growth of her manufacturing and pastoral pursuits with those of any other country, say America, we find that the former have by no means kept pace with the latter; and although the manufacturing industries of Australia have undoubtedly increased and multiplied in a manner most gratifying to all patriots, they cannot yet be even momentarily compared to the pastoral interests of the country. Since the year 1810, when only 167 lbs. of wool were exported, up to the year 1879, when 287,757,934 lbs. of wool found their way to foreign markets, the large increase of wool grown, and capital invested in this business, are only to be compared to the rapid increase of manufactures of all sorts in North America ; and, from being a comparatively un- known place, Australia has, in a short time, achieved the proud position of the greatest wool-supplier of the world. Despite many great disadvantages, some natural and others artificial, such as the great distance from foreign markets, and the introduction of convicts years ago, Australia bids fair to retain this position, and to increase her lead over foreign competitors, the most formidable of whom is South Africa. The importance of the wool-producing business, and the vastness of its extent, are indeed difficult to realize; and when we look ahead of us a few years, the immense capabilities which this industry is almost certain to develop become too great and magnificent for human realization.

We cannot, of course, look forward to as great an increase in the productive powers of the colonies during the coming seventy years as we are now able to look back upon, and it is perhaps as well that this is the case, for if it were not the supply would undoubtedly exceed the demand, and so check the prosperity of the whole of the colonies. But we can certainly count upon a large and gratifying increase.

Year by year our back country is being taken up, and its resources developed, by enterprising men, who will doubtless achieve that success they so richly merit; and instead of the character of the country proving to be what many "birds of ill-omen" had predicted—dry, arid, and waterless, incapable of growing grass, and quite unfit for pastoral or agricultural tenure—it has been proved to be rich and varied pasture lands, abounding in salt, cotton, and other bushes upon which stock luxuriate and fatten, and to have numerous creeks with good water-sheds, and every natural facility for tank or well-sinking and dam-making. And although droughts have been experienced, it cannot be stated that the back country has suffered in any greater degree than that situated nearer the coast. It has been found, however, that creeks which for many years had been unknown " to run " did so frequently after the introduction of dams and stock.

Winters seldom pass now without a rich water supply being received in the back blocks, and it is indeed rare to pass the time from February to August without having the pleasure of a large or only partial flood.

Another good quality possessed by the back country is that of resuscitation, if we so term it. By this we mean that quality which enables country that has been over-stocked and stripped of every vestige of verdure to assume, in an incredibly short period of time, when relieved of its stock, a thick and luxuriant growth of grass, and to present once more all the appearances of unartificial pasture land of the richest description.

Not only is the grass of many parts of Australia much more luxuriant than the description of soil would lead one to expect, but the soil itself, presenting as it does an appearance of sandiness considered by many incompatible with rich growth, is naturally capable of producing good crops, and, if aided by artificial means, may, and will, be developed into arable lands capable of producing crops unsurpassed in any other part of the world.

To the discovery of gold, and consequent influx of population and capital, is to be ascribed much of the prosperity of the colonies ; but to the energy, enterprise, and hardihood of many of our old colonists the credit is mainly due, and it is to these qualities, and to the natural advantages of the country, that we have to look for the causes of the great changes that have taken place, and them that we have to thank for rendering a country, not many years ago considered good enough for nothing but a penal settlement, one of the grandest of the British possessions, and one of the most promising portions of God's fair earth.

The continent of Australia, for so it may fairly be called, is divided into five colonies, viz.—New South Wales, Victoria, Western Australia, South Australia, and Queensland, the respective areas of which are—

New South Wales ...	323,437 sq. m.	206,999,680	acres
Victoria	88,198 ,,	56,446,720	,,
Western Australia ...	978,298 ,,	626,111,323	,,
South Australia ...	914,730 ,,	585,427,200	,,
Queensland	678,600 ,,	434,304,000	,,
Total area ...	2,983,263 sq. m.	1,909,288,923	acres.

In 1880 the number of sheep depastured in each of the colonies was as follows:—

New South Wales	...	...	...	32,399,547
Victoria	...	...	...	10,355,282
Western Australia	...	...	...	1,231,717
South Australia	...	...	...	6,463,897
Queensland	...	...	...	6,935,967
Tasmania	...	...	...	1,783,611
New Zealand ...	...	...	...	13,069,338

By this table we see that the colony containing the largest area, namely, Western Australia, has considerably less than one-twentieth part of the number of sheep depastured in New South Wales, although the latter contains less than one-third of the area of the former.

This disparity is to be accounted for in many ways. Among others may be mentioned the uninviting aspect of the physical features of a great part of the colony, and the great scarcity of labour, which had become so great as to threaten the entire emigration of the population, until, in 1851, at the request of the colonists, convict labour was introduced. Although in the western portion of the colony a large desert exists, which renders the country almost unavailable, there are many hundreds of thousands of acres of splendid pastoral country, suitable for squatting, and which we feel certain will, before many years have elapsed, be divided into sheep and cattle runs. An abundance of water can be obtained, where watersheds and creeks do not exist, by sinking wells. The land laws of Western Australia are very liberal.

NEW SOUTH WALES is, in regard to the number of sheep depastured, by far the most important of the colonies, and exports annually, in round numbers, 76,000,000 lbs. of wool, to the value of about £5,025,000.

VICTORIA.—This colony, which has the smallest area of any of the five we are treating upon, is only about one-quarter of the size of New South Wales, and contains half as many sheep, and fully twice as many as any of the other colonies, except New Zealand.

The sheep bred in Victoria are of a larger and better description than those of the other colonies. They enjoy more attention than do the others, and are more carefully bred. With few exceptions, Victorians pay more attention to the quality than the number of their sheep, which we are sorry to see is not the case with the majority of New South Wales squatters. At the Paris exhibition, however, the first prize for fine wool was awarded to Mr. Cox, which proves that some of the New South Wales squatters are capable wool-growers; but, taking the whole production of the colonies, Victoria is ahead both in sheep and in wool, and also in the average of wool per head.

Unfortunately, the policy of the Victorian Government regarding border duties on stock has had a most deteriorating effect upon the meat trade of Victoria and Riverina, which has made itself severely felt by the squatters of Riverina, who now find it more to their advantage to sell their stock in Sydney rather than the Victorian markets.

From the statistics of Mr. Hayter, the Government statist, we quote the following regarding the number of sheep depastured in Victoria from the year 1863 to 1874 inclusive :—

1863	...	...	7,115,943	1869	...	...	9,923,663
1864	...	...	8,406,234	1870	...	...	10,761,887
1865	...	...	8,835,380	1871	...	...	10,002,381
1866	...	...	8,833,139	1872	...	...	10,575,219
1867	...	...	9,532,811	1873	...	...	11,323,080
1868	...	...	9,756,819	1874	...	...	11,225,206

SOUTH AUSTRALIA depastured in 1874 about 6,000,000 sheep. Since that time, however, long and severe droughts have visited some parts of the colony, proving very destructive to stock, notwithstanding which, in 1879 the number depastured was about 6,377,812 sheep.

QUEENSLAND has of late come into much notice from squatters, many of whom are investing their capital in station properties there; a portion of them having been driven from Victoria by the misguided policy of the Government. The land is well suited to the successful following of pastoral pursuits, and the future of this colony is full of promise.

THE COLONY OF TASMANIA, although separate from Australia, forms part of Australasia. The length from east to west is 200 miles, breadth from north to south 185 miles, and it contains 24,330 square miles; the different islands, 1,885 square miles. Total area, 26,215 square miles, or 16,777,600 acres. It was, in 1642, discovered by A. J. Tasman, and now bears his name, although christened by him Van Dieman's Land. The character of the country is in general hilly, and Tasmania possesses some very high mountains, varying in height from 1,700 to 5,000 feet, the highest being Ben Lomond, which rises to a height of 5,002 feet, and the lowest Bruni, 1,660 feet. Tasmania was at one time evidently a portion of the continent, as is seen by the range of granite mountains extending across the Straits in the direction of Wilson's Promontory. Among Tasmanian breeders are some who can compete successfully with any in the Australian colonies, both with their sheep and horses; " Scotch Jock " being almost a household word in the colony.

NEW ZEALAND.—These islands are almost the anti-

podes of the British Isles. They are named respectively
North, South (sometimes Middle), and Stewart's Island.
They are situated about 6,500 miles west from the coast
of South America, and about 1,200 miles south-east of
Australia. The dimensions and areas of the islands are
as follow :—

> North Island is 500 miles long, 200 broad, and
> contains 48,700 square miles, or 31,168,000
> acres.
> South Island is 550 miles long, 210 broad, and
> contains 72,100 square miles, or 46,144,000
> acres.
> Stewart's Island is triangular in shape, and contains
> 1,800 square miles, or 1,152,000 acres.
> Total, 122,600 square miles, or 78,464,000 acres.

Mount Cook, in the South, or Middle Island, is the
highest mountain in New Zealand, and rises to a height
of 14,000 feet above the sea level.

The number of sheep depastured in 1880 was
13,069,338, and the quantity of wool exported in the
same year was, in round numbers, 60,000,000 lbs., valued
at about £4,000,000; as against, in 1875, 11,754,621
sheep, yielding 54,401,540 lbs. of wool, valued at
£3,398,555; and these figures show an increase over the
corresponding ones of 1874 of, in lbs., 7,553,540, and in
value, £563,860.

The total wool clip of the Australian colonies was, in
1877, in round numbers, 284,000,000 lbs. This, when
scoured, gives us 168,259,098 lbs., having lost in the
process of scouring 115,741,002 lbs., or a percentage of

41 and a small fraction. Having carefully worked out these figures, we may say that we consider them to represent fairly the loss in weight generally sustained through scouring. We have heard it asserted that wool will sometimes lose as much as 60 per cent. of its weight during scouring or washing, but this we cannot verify from personal experience (speaking of fleece wool, or even of an entire clip), although we consider it quite probable that in some places, such as the sandy districts of the Darling (where we have seen a sheep-proof yard, that has been built in the morning, rendered almost useless before night, from the quantities of fine sand that had been blowing in clouds all day), this might be the case.

Further information regarding the percentage of weight lost by washing and scouring wool, and the comparative advantages of the various methods adopted, will be found under a separate heading, where the subject is more fully treated upon.

The wool clip of Buenos Ayres and the River Plate for 1877 was 222,500,000 lbs., and that of other countries, including South Africa and America, was for the same year (1877) 463,000,000. The total wool clip of the world for 1877 was about 1,497,500,000 lbs., worth about £30,000,000. This, when scoured, would give a yield of about 852,000,000 lbs. of clean wool. The production of California in 1877 was 53,110,742 lbs. as against 56,550,970 lbs. in 1876.

CHAPTER II.

VARIETIES OF SHEEP.

ALL the wild sheep known are natives either of mountainous regions or of dry and elevated table-lands. They are gregarious, a character which the domestic sheep fully retains. They are generally seen in small flocks, and are not easily approached, taking refuge in flight, a sharp, whistling sound emitted by one of the rams serving as an alarm to the whole flock; although they are very capable of making a vigorous defence when driven to close combat. A ram of domestic species is, indeed, able to sustain a contest with a bull, taking advantage of his greater agility, and butting against his foe with his strongly armed forehead. A ram has been known to throw a bull on the ground at the first onset, and is always ready to defend himself and his companions against a dog. Many rams exhibit great pugnacity. Sheep differ from goats in their mode of fighting. Goats rear themselves on their hind legs and throw themselves sideways on their adversaries, to bring the points of their horns to bear. Sheep rush straight at each other, a mode which better suits the different style of armature of the head. Rams of the black-face variety are especially powerful with their heads, and often, at the rutting season, kill each other. Their naturally strong skull is considerably protected in battle by heavy arched horns. A thorough ram fight is a terrifying sight. The two warriors go backwards each some fifteen or twenty yards, and they meet each other with great violence, their heads cracking loudly, and their beam-ends rising in response to the collision of heads. Ewes of this breed fight also. Sheep without horns are not so pugnacious as the mountain breeds. All the wild sheep have short wool, with an outer coating of long and nearly straight hair. But even the long hair—at least on the moufflon—has the peculiar character of wool, in that roughness of surface which gives it the property of "felting." One effect of domestication in the common sheep has been to cause the disappearance of the outer long hair, and to produce instead an increase in the length and abundance of the wool, an object of great importance to the sheep-farmer.

In neglected breeds of the common sheep, the two kinds of hair or wool are very apparent. In some tropical climates the sheep loses its abundant fleece, and is covered with hair little longer than that of the ox.

A very interesting species of the wild sheep is the ROCKY MOUNTAIN sheep, or BIG-HORN, of North America. It is equal in size to the argali, which it much resembles also in its general appearance, and in the size and curvature of its horns.

The horns of the old rams attain so great a size, and are so much curved downwards and forward, that they often effectually prevent the animal from feeding on level ground. The abode of this species is in the most craggy and inaccessible parts of the Rocky Mountains. The flesh is of the very finest quality. The wool is very fine, and fully one inch and a half long; it is completely concealed by long hair. The general colour is brown, paler on the lower parts; the old rams are almost white in spring.

The AOUDAD is a native of the north of Africa, inhabiting chiefly the lofty part of the Atlas Mountains. It is sometimes called the BEARDED ARGALI, although it has no beard on the chin; but the throat, the chest, and the front of the fore legs are remarkably adorned with long, shaggy hair. On other parts the hair is comparatively short, with an underclothing of short wool. The colour is a uniform reddish-yellow. The tail is longer than on the other wild species, and is terminated by a kind of tuft of long hair. The horns are not so large as in the other wild species. In size the aoudad exceeds the moufflon, but is not equal to the argali. The French call it *Moufflon à Mouchettes*, or Ruffled Moufflon, from the long hair on its legs.

The common sheep was probably the first animal domesticated by man. The leather made of the skin of the sheep is much employed in bookbinding and for making gloves. In patriarchal times the milk was much used, as it still is in some countries; it is richer than cows' milk, and the cheese made of it has a sharp taste and strong flavour, which, however, are greatly relished by some. In Britain the milk is now very little used.

In some mountainous parts of India the sheep is even used as a beast of burden, carrying loads of from 35 lbs. to 40 lbs. over rough tracks and up steep crags, where almost no other animal could be employed. Those who watch sheep carefully, or keep them as pets, find them by no means devoid of intelligence. They have, however, a stupid habit of following, without scruple, the leaders of the flock; so that, when sheep are being driven across a narrow bridge, or where a fence separates the road from a precipice, if anything occur to deter them from proceeding on the proper path, and one break over the fence or parapet, more of the flock may be expected to follow, as has sometimes happened, to their utter destruction. Sheep very soon come to know the voice of the shepherd, and also the appearance as well as the bark of the shepherd's dog. Though they stand more in awe of the shepherd's voice, or commands, than of any other human being's, the dogs regularly moving amongst them fail to keep them in such subjection as strange ones do.

The breeds of sheep are very numerous and very different. The BLACK-FACED sheep of the Highlands of Scotland and of the north of England is perhaps as near the original type as any existing breed. Both male and female have horns; those of the ram large, with two or more spiral twists; those of the ewe much smaller, and little twisted. The face and legs are not always black; many are speckled, and some principally white. The black-faced sheep is robust, very active and hardy, enduring the rigour of a severe winter when sheep of most of the breeds common in Britain would perish. It survives on little food, and shifts admirably for itself in a snowstorm. The small quantity and even inferior quality of food with which a black-faced sheep will tide over a snowstorm is most surprising. As an instance of the tenacity of life in black-faced sheep, under certain circumstances, they have been known to be buried five weeks under a snow-wreath and come out alive. It has a bright, quick eye, with an expression very different from that softness which is seen in many of the breeds preferred for lower grounds and better pastures. The wool is long and coarse, and the weight of the fleece from 3 lbs. to 4 lbs. ; but the mutton is of the finest quality, and on this account, and its hardiness, this breed is preferred to any other in many mountainous districts, and on rough elevated moors.

The WELSH sheep is much smaller than the black-faced; both sexes horned; the colour various; the mutton highly esteemed; the fleece seldom weighs 2 lbs.

A very little larger breed, with big bushy tail, hornless, or with short and little twisted horns, has long existed in the Shetland and Orkney Islands, its wool affording the material for the manufacture of Shetland hose. The Shetland and Orkney sheep are very hardy, and in winter feed much on sea-weed.

Smaller than either of these, and, indeed, remarkably diminutive, is the hornless BRETON sheep.

The forest sheep of England, so called from being pastured in the royal forests, have now in most places been supplanted by other breeds. They are still to be seen on the barren grounds between the British and Bristol channels; and the mutton is in much request in the London markets. The original forest sheep was generally small, with face and legs russet brown or grey; wild, restless, and difficult to fatten, but producing wool of fine quality. The DORSET sheep is one of the best of the old English upland breeds. Both sexes have small horns. The wool and mutton are of medium quality, but the ewes are remarkable for their fecundity and the abundance of their milk; and this breed is valued as affording a supply of early lamb for the London market.

The RYELAND sheep has long existed in Herefordshire and some neighbouring counties in England. It is small, short-limbed, white, hornless, produces excellent mutton; and before the introduction of merino wool, this wool was preferred to every other kind for the manufacture of the finest broadcloths.

The CHEVIOT sheep has existed from time immemorial on Cheviot Hills, and is now very widely diffused over a considerable part of England, and almost all parts of Scotland, being hardy, and well adapted for high grounds, although it is inferior in hardiness to the black-faced. Cheviots, however, rather excel the black-faced, both in size and in the value of their fleeces, but require a richer pasture. Ewes are hornless, and the rams almost so. The general figure is longer than that of the black-faced sheep. They are narrow in shape, with splendid forequarters, and long pricked ears. The colour is white, the face and legs occasionally mottled with grey, but generally quite white. The fleece weighs from 3 lbs. to 5 lbs. Great attention has for many years been devoted to the improvement of this breed.

The LEICESTER is another of the most valuable breeds. This breed, as it now exists, as a result of the skill and care of Mr. Bakewell, who, soon after the middle of last century, began to make experiments for the improvement of the old Leicester sheep—a large, coarse-boned sheep, not easily fattened, and with coarse, long wool, of which, however, the fleece weighed from 8 lbs. to 10 lbs. The new Leicester sheep has wool moderately long, of better quality, the average weight of the fleece being about 7 lbs. or 8 lbs., and is easily rendered very fat. It is naturally very broad in the back, with finely-arched ribs. The colour is white. Both sexes are hornless. The Leicester sheep are now common in all but the mountainous parts of Britain, and other breeds have been improved by crossing with it, particularly various breeds of long-woolled sheep, which have long existed in different parts of England—as those of Lincolnshire, Romney Marsh, &c.

A famous long-woolled sheep is that called the COTSWOLD, or GLOUCESTER, the wool of which was in great esteem in the fourteenth and fifteenth centuries, bearing a higher price than any other wool. In 1464 Edward IV. sent a present of Cotswold rams to Henry of Castile, and in 1468 a similar present was sent to John of Arragon. The Cotswold breed, however, as it at present exists, has been modified by crossing with the Leicester, and produces better mutton, and shorter wool, than in former times.

The SOUTHDOWN sheep has recently been improved with the utmost care. The colour is generally white, and the face and legs are generally dun, black, or speckled. Both sexes are hornless. The wool is short, very close, and curled. The Southdown derives its origin and name from the chalky downs of the south of England; but is now met throughout England and the south of Scotland.

The SHROPSHIRE sheep are large, with thick wool, something like the southdown. They are hornless, and black or dun in the face and legs. They come early to maturity, but are suited only for finer climates and good keeping.

The OXFORD-DOWN is a heavy, somewhat soft sheep, without horns, and capable of rapid and great development under good treatment. It is not suited to very cold and exposed positions.

The ICELAND sheep is remarkable for very frequently having three, four, or five horns. They are good butchers' animals, being deep and thick in the carcase, though rather short on the quarter. The same peculiarity, or monstrosity, as it may be deemed, is exhibited by the sheep of some of the most northern parts of Russia.

The north of Africa possesses a breed of sheep with legs of great length, pendulous ears, and much arched face; the wool short and curled, except on the neck and shoulders, which have a kind of mane. India has also a hornless breed, with pendulous ears, short tail, and very fine, much curled wool. The broad-tailed or fat-tailed sheep is found in many parts of Asia, as in Syria, India, and China, also in Barbary, and is now very abundant in the colony of the Cape of Good Hope. It is of rather small size, with soft and short wool. Its chief charac-

3

teristic is the enormous development of the tail, by the accumulation of a mass of fat on each side, so great that the tail has been known to weigh seventy or eighty pounds. The tail is highly esteemed as a delicacy, and, to protect it from being injured by dragging on the ground, the shepherd sometimes attaches a board to it, or even a small carriage with wheels. The fat of the tail is often used instead of butter; it is less solid than other fat. The fat-rumped sheep of southern Tartary has a similar accumulation of fat on the rump, falling down in two great masses behind, and often entirely concealing the short tail.

The ASTRAKHAN, or BUCHARIAN, sheep has the wool twisted in spiral curls, and of very fine quality.

The CIRCASSIAN sheep has a remarkably long tail, covered with fine long wool, which trails on the ground.

The WALLACHIAN sheep, common in Hungary, as well as in the country from which it derives its name, is distinguished by the magnitude of its horns, and their direction. They make one great spiral turn, and then generally rise up from the head to a great height, twisting round as they rise. The wool is soft, and is concealed by long hair.

MERINO SHEEP.—There are now many varieties of the merino, all, however, descended from the old Spanish flocks, which have acquired distinctive appellations, such as the French, Saxon, Silesian, Leonese, Paular, American, Australian, and other merinos.

THE SPANISH MERINO.—The origin of this animal is involved in obscurity. The commonly received account is that Columella, a Roman who resided near Cadiz in the reign of Claudius, coupled fine-woolled Tarentian (Italian) ewes with wild rams brought from Barbary, and thus laid the foundation of the breed; that, some thirteen centuries after, Pedro IV., of Castile, improved it by a fresh importation with rams from the same country; and that, two hundred years later still, Cardinal Ximenes a third time repeated this ameliorating cross; from which period, we are left to infer, the breed became established about as it was found when it first began to attract the special attention of foreign nations, in the seventeenth century. All the early varieties of Africa had long, straight, hairy wool, like the present long-woolled sheep of England; and no writer, ancient or modern, has pretended that the rams imported from that country into Spain were any different in this particular. How recurring crosses between such animals and fine-woolled ewes should have commenced, improved, and finally fixed the characteristics of a breed like the merino is a problem which admits of no rational solution to a practical sheep-breeder.

Strabo, who was a contemporary of our Saviour, and who consequently lived a generation earlier than Columella, says that the fine cloths worn by the Romans in his time were manufactured from wool brought from Truditania, in Spain. Pliny, himself Governor of Spain, writing after Columella's time, describes several fine-woolled varieties in that country which must have existed there a long time anterior to Columella. The Barbary crosses undoubtedly were made with or formed the

Chumah, or long-woolled breed of Spain, which is altogether distinct from the merino. This pedigree is
probably entitled to about as much confidence as that
which the Greek poets gave to the wonderful ram which
bore the "Golden Fleece." He, according to this very
respectable authority, was got by the sea-god Neptune,
dam the nymph Theophane.

The only well-settled facts on this subject—and for-
tunately they are sufficient for all practical purposes—
are, that at a period anterior to the Christian era fine-
woolled sheep abounded in Spain ; that they were
preserved, and made themselves heard of in the channels
of trade and the domestic arts through all the conquests,
reconquests, and other sanguinary convulsions of that
kingdom ; that they were, or gradually ripened into, an
exclusive breed, unique in its characteristics, and essentially unlike all other breeds in the world.

When the merinos of Spain first attracted the observation of other nations they were found scattered over
most portions of their native country, divided into pro-
vincial varieties which exhibited considerable differences ;
and these again were separated into great permanent
flocks, or cabanas, as the Spaniards termed them, which
had so long been kept distinct from each other, and subjected to special lines of breeding, that they had acquired
the character of sub-varieties, or families.

The merinos, as they appeared as a race at the beginning of this century, are thus described by Livingstone,
in his valuable " Essay on Sheep," which is a recognized
authority throughout the world :—

"The race varies greatly in size and beauty in different parts of Spain. It is commonly rather smaller than the middle-sized sheep of America. The body is compact, the legs short, the head long, the forehead arched. The ram generally (but not invariably) carries very large spiral horns, has a fine eye and a bold step. The ewes have generally no horns. The wool of these sheep is so much finer and softer than the common wool as to bear no sort of comparison with it; it is twisted and drawn together like a corkscrew; its length is generally about three inches, but when drawn out it will stretch to nearly double that length. Though the wool is, when cleaned, extremely white, yet on the sheep it appears a yellowish or dirty-brown colour, owing to the closeness of the coat, and the condensation of the perspiration on the extremities of the fleece. The wool commonly covers a great part of the head, and descends to the hoof of the hind feet, particularly in young sheep; and it is also much more greasy than the wool of other sheep."

We select the following from a more extensive table by Petri, published in the early part of this century. It may be found of use, as well as interesting to those who wish to make practical comparisons between these sheep and their descendants.

Names of Flocks.	Weight, including wool.	Length from mouth to horns.	Length from horns to shoulders.	Length from shoulders to tail.	The whole length.	Circumference of the belly.	Height of the fore legs.	Height of the hind legs.	Distance of hip bones apart.
Negretti	*lbs.*	*in.*	*ft. in.*	*ft. in.*	*ft. in.*	*ft. in.*	*ft. in.*	*in.*	*in.*
Ram	97	9½	1 7	2 2	4 6½	4 1½	1 3	10	6
Ewe	67	8½	1 5	2 1	4 2½	4 1½	1 1	9½	4½
Infantado									
Ram	100½	10	1 6	2 3	4 7	4 2	1 0	9	6
Ewe	70	9	1 5½	2 1	4 3½	3 11	1 0	8½	5½
Guadeloupe									
Ram	97½	9	1 6	2 2	4 5	4 5½	1 0	8	6
Ewe	69	9	1 2	2 1	3 11	3 9	0 10½	6½	4
Estantes of Sierra de Limo									
Ram	96½	9½	1 6	2 0	4 3½	4 2½	1 0	8	6
Ewe	62½	9	1 2	2 1	4 0	3 10	0 11	7	5
Small Estantes									
Ram	42	7½	1 3	1 9	3 7½	3 2	0 10	6½	3
Ewe	30	7	1 1	1 6	3 2	2 10	0 8	6	3

Gilbert, in his report to the National Institute of France, in 1796, thus describes the sheep imported into France by Louis XVIII., with the permission and assistance of the King of Spain, of which about 300 arrived safely in 1786:—

"The stock from which the flock of Rambouillet was derived was composed of individuals beautiful beyond any that had ever before been brought from Spain; but, having been chosen from a great number of flocks in different parts of the kingdom, they were distinguished by very striking local differences, which formed a medley disagreeable to the eye, but immaterial as it affected their quality. These characteristic differences have melted into each other by their successive alliances, and from thence has resulted a race which perhaps resembles none of those which composed the primitive stock, but which certainly does not yield in any circumstance to the most beautiful in point of size, form, and strength, or in the fineness, length, softness, strength, and abundance of fleece.

"The comparison I have made, with the most scrupulous attention, between this wool and the highest-priced of that drawn from Spain, authorizes me to declare that of Rambouillet superior."

In reference to this Mr. Randall says:—" Judging by the taste uniformly displayed by the French in that particular, there is little doubt the 'abundance of fleece' was the first rather than the last consideration—as it here happens to be named—which guided the original selection. And the far more liberal feed which the sheep received in France, the exemption from the exhaustive annual migrations of Spain, and a course of breeding specially designed to produce that result, rapidly carried the weight of their fleeces beyond any point ever known in their native country.

"Ten years after their introduction into France, Lasteyrie gives their average weight of fleeces, unwashed, and thus continues it through a series of years: in 1796, 6 lbs. 9 oz.; 1797, 8 lbs.; 1798, 7 lbs.; 1799, 8 lbs.; 1800, 8 lbs.; and 1801, 9 lbs. 1 oz."

Gilbert, writing under Government patronage, said, in 1796:—" Almost all the fleeces of the rams, of two years old and upwards, weigh from 12 lbs. to 13 lbs., but the mean weight, taking rams and ewes together, has not quite attained to 8 lbs., after deducting the tags and the wool from the belly, which are sold separately.

" Leaping over a chasm of twenty-five years, let us again examine the Rambouillet sheep, and ascertain the progress of this most interesting experiment, through the eyes of an English breeder of merinos.

" Mr. Trimmer, the author of the ' Practical Observations,' visited this flock in 1827, and the following is his often quoted description of it :—' The sheep in size are certainly the largest pure merinos I have ever seen. The wool is of various qualities, many sheep carrying very fine fleeces, others middling, and some rather indifferent ; but the whole is much improved from the quality of the original Spanish merino. In carcase and appearance I hesitate not to say they are the most unsightly flock of the kind I ever met with. The Spaniards entertained an opinion that a looseness of skin under the throat and other parts contributed to the increase of fleece. This system the French have so much enlarged on that they have produced in this flock individuals with dewlaps almost down to their knees, and folds of skin on the neck like frills, covering, nearly, the head. Several of these animals seemed to possess pelts of such looseness of size that one skin would nearly hold the carcases of two such sheep. The pelts are particularly thick, which is unusual in the merino sheep. The rams' fleeces were stated at 14 lbs., and the ewes' 10 lbs., in the grease. By washing they would be reduced half, thus giving 7 lbs. and 5 lbs. each.' "

In writing to Mr. Randall, Mr. Taintor says :—

"In 1828 I imported a lot of Saxony sheep, and at various times have selected in France nearly one thousand of their best merinos. In 1842 my friend, D. C. Collins, of this city (Hartford), bought, by my advice, fourteen ewes and two rams of the royal flock at Rambouillet. About half of them were good sheep, but, from want of care and attention, the importation was of but little value to the owner or the country.

"I cannot afford to keep any other sheep for wool but French merinos. I call them best because they pay best, and that is the true test. Not the sheep that can crawl through the year with the least possible care and feed, but are generously cared for and bred with close attention and judgment, with always an eye for the most valuable fleece for the manufacturer and the most valuable carcase for the butcher.

"Since 1828 I have been seven times across the water, and at one time took a year and a half to go to every part of Europe and examine the flocks and see the owners, hear all they had to say, and then use my own judgment. You are aware that the Spanish merinos have become almost lost. They are so small, neglected, and miserable that I would not take one of them even as a present.

"Improved machinery, too, has had a ruinous effect upon the Saxony flocks, as they have learned the art of using medium wool in the place of very fine.

"The sheep of Saxony proper are more than half a million less in number than they were ten years ago.

" In France the royal flock, now the property of the Emperor, at Rambouillet, which for years attracted all the sheep-masters to its annual auction sale, bred the fleece so fine and the animal so delicate that they could no longer attract attention; and, four years ago, they changed the plan, and now sell (when they can) at private sale. The sheep have no wool on the head or legs, and but little on the belly; they are ruined by high breeding. The wool is short and fine.

" Ewes' fleeces average 14 lbs. (in flocks of 500), and rams' 20 lbs. to 24 lbs. Average weight of ewes, say (all ages) 100 lbs., and rams 200 lbs. One ram I bought (for 3,000 francs, or £125) weighing 309 lbs., carring a fleece, unwashed, of 32 lbs. Fair estimate of loss in washing, 60 per cent.

" It is from this class of flock I have selected my merinos. It is from wool of this class that the fine French 'muslins de laines' are made, as it has length of staple and fineness, with requisite strength, which is all-important.

" Three years ago a gentleman sent me, from Estremadura, a number of Spanish merino fleeces as a sample (as circumstances did not allow me to see the flock when in Spain); they are little wads of fleeces."

The only weak point of the best French merino, as a wool-producing animal, is the want of that hardiness which adapts it to our changeable climate.

The French sheep has not only been highly kept and housed from storm and dew for generations, but it has been bred away from the nominal type of its race. The

Dishly sheep of Mr. Bakewell are not a more artificial variety, and all highly artificial varieties become comparatively delicate in constitution.

The following frank and well-condensed opinions on this subject are from the pen of an American writer, Mr. F. M. Rotch, who imported to his country, in conjunction with Mr. Taintor, some of these sheep, and who, a few years ago, had a most admirable flock of them. He writes :—

"France I visited two or three times, with a view to import merinos, and sent out to Mr. Taintor quite a number of the French variety. The French merinos of the first class are certainly superb sheep, but they vary there as they do here—a few flocks, say half a dozen, being very superior, and then come a number of mediocre flocks, where neither care, expense, nor knowledge are bestowed, and where the sheep more closely resemble the old Spanish type. You ask my opinion of the French, as suited to rough farming. I don't think them at all fitted to it. Though a vigorous, good-constitutioned, and hardy sheep, they are accustomed to too much care and watchfulness in their native land to be able to endure the rough-and-tumble style of much of our farming. The north side of a barn and the lee side of a rail fence, for animals that are housed every night in the year at home, is too sudden and great a change. With proper care they are able to endure even our vicissitudes of climate, and thrive and grow fat here as in France ; but, like all improved breeds of domestic animals, it is folly to expect them to do well without care

and feeding. Any animal brought from a state of
high cultivation and a mild temperature to a colder
climate and poorer soil will deteriorate unless extra
pains are taken to supply the loss of care and coun-
teract the change of food. During the dozen years I kept
French merinos, I gave them much the same care they
had in their native country, and found them to thrive,
and breed, and weigh, and shear as they did there, *almost.*
They are good breeders and nursers, often having twins,
and rearing them well. As a cross upon our usual type
of merino I consider them very valuable, but quite unfit
for general use as a stock sheep by our farmers at
present."

In these able remarks upon the French merino, and
impartial criticisms, we quite coincide, and consider that
they are little suited for general breeding in Australia.
They, from the long course of careful breeding, and from
the system of housing and feeding them, have become so
delicate in constitution as to be quite unfit for turning
out in the large paddocks of most of our Australian sheep
runs. The French merino, then, as it now appears, lacks
altogether the hardiness and endurance of the Spanish
sheep, and it is to this sheep that we will now attract
the reader's attention.

The Spanish merino now occupies the most prominent
position as a wool-producing sheep, and is especially
suited to Australia, where the growth of fine wool and
large fleeces is more an object than the production of
mutton. In this respect it is the merino sheep is
deservedly a favourite among Australian breeders, and is

bred by them in preference to the Leicester, Lincoln, and other large-framed sheep, whose growth of wool is greater but not so valuable, but whose qualities as mutton-producers are far superior to those of the merino; and it is for this reason that the merino is almost unknown in England, and large-framed sheep are bred, because that there mutton is the great object of the sheep-farmer.

Another quality in the merino to recommend it to Australian sheep farmers is its hardiness, for, although fattening more slowly, the merino will thrive on less food and water than any other sheep, and this is a great object to the breeder in Australia, where the runs are so large, and the water mostly artificial, and at great distances.

The first sheep introduced into Australia were brought in January, 1788, and numbered 29; but the first merino sheep of which we have any record were introduced by Mr. John M'Arthur, from the Cape of Good Hope, and consisted of one ram and five ewes. They arrived in 1796.

These were crossed with Mr. M'Arthur's original flock of Bengal sheep, and the favourable results soon became apparent, in the conversion of the hairy coat into wool, which was also assisted by very favourable climatic influences. The sum of £300 for one pure-bred merino ram, or £500 for a pair, the progeny of the pure Spanish merino introduced by Mr. M'Arthur, was offered at Paramatta in 1822.

To show the rapid increase in the numbers of these

sheep we may mention that, 21 years later, in 1843, the process of boiling down was commenced at Yass, by Mr. O'Brien, and sheep which were otherwise not worth 2s. 6d. per head, produced, by this process, from 5s. to 8s. per head; and from this date the export of tallow became an important item in trade.

A further introduction of merino blood was made in 1806, by Mr. M'Arthur, who, with the assistance and encouragement of the Privy Council, selected from the flock of George III. three rams and two ewes for this country.

CHAPTER III.

BREEDING.

In a great measure the rules which guide the breeding of stock have been learned by experience, and are rather to be regarded as contributions to science than as deductions from it. The probable relative influence of the male and female parent upon their progeny is a point unquestionably of the greatest importance, but concerning which widely different opinions have been maintained; and another much controverted and important point is, the propriety of *breeding in and in*.

Practically, the rule is always observed by those who seek the improvement of a *breed*, of selecting the very finest animals possible, both male and female, although a great improvement of the existing stock on a farm or station is often effected in the most advantageous manner by the mere introduction of males of better quality. The dangers of breeding *in and in* are very generally acknowledged, even whilst it is contended that they may be very much obviated by the careful rejection of every faulty animal, and that in this way the utmost advantage may be taken of the very highest improvements ; but it is likewise very generally admitted that, if equally improved individuals can be obtained not so nearly related, it is better to seek the perpetuation of the breed by their means. It is a rule also of much practical importance, that an improvement of breed is to be attained not by a *cross* between animals of very different breeds, as between a dray-horse and a race-horse, but only between those which are comparatively similar. The result of the intermixture of very dissimilar breeds is never in any respect satisfactory. Upon this subject we quote the following from Mr. H. S. Randall :—

" The art of breeding is the art of selecting and coupling those males and females which are best adapted to produce an improved and uniform offspring. The first great rule of breeding is that like produces like. But this must be held to extend to blood as well as individual characteristics, or else it is a rule which will mislead the inexperienced.

" In selecting animals for coupling, especial pains should be taken not to interbreed those possessing the *same* defect, because in that case observation proves that the offspring inherit something like the aggregate of the defect of both parents—that is to say, if the ram is defective in the crops (in proper fulness back of the shoulders) to an extent expressed by 2, and the ewe to an extent expressed by 3, their offspring will possess the defect to something like the extent of 5. Of course, this rule is not invariable, and would not continue to apply to its full extent if breeding between the produce of those similarly defective animals was continued, for in that case they would soon have no crops at all.

"A defect may be an individual or family one. The latter is far more likely to be transmitted to the progeny. The other sometimes appears to be accidental, and is not forcibly transmitted. I would rather breed from a slightly defective animal from a very perfect family than from a very perfect animal from a slightly defective family.

"It would be strictly accurate to say that if animals possessing the same defect are interbred with each other, the offspring should be *expected* to inherit that defect to a greater extent than either parent, and that continuing such a course of breeding would soon increase the defect to the greatest practicable extent, and in the case of defects affecting the constitution of the animal, to a fatal extent.

"The defects of one parent should be met by peculiar excellence of the other parent in the same point. If a dam is 'high on legs,' she should be bred to a ram with short legs; if thin-fleeced, to an uncommonly thick-fleeced ram, and so on. This, however, is to be understood within certain limitations. These counteractions are to be sought within the circle of proper excellence and proper uniformity in other particulars. The distinguishing features aimed at in the flock are neither to be sacrificed nor constantly changed or disturbed for the purpose of producing a sudden amendment in a single point.

"There is a practical fact of the utmost importance in the selection of breeding rams. All do not transmit their qualities in an equal degree to their offspring. The power to 'mark offspring,' as it is termed, depends most on two properties. The first and by far the most influential of these is blood. By blood I mean nothing mysterious or unexplainable. I simply mean that blood which has flowed so long in one distinct channel, and through animals so closely alike in all their properties, that it has acquired a power resembling that of species —a power continuously to reproduce animals of the same family, and almost the same individual characteristics. Under this definition the unsightly ass may have as high and as pure blood as the winged courser of Arabia—the miserable, hairy, broad-tailed sheep of Asia and Africa as the far-descended merino of Spain.

"The ram should not only, then, have a faultless pedigree, but, if practicable, be drawn from an old, distinct, well-marked family of merinos that have been the same as a whole, and uniform among themselves, for a long course of generations. I used to notice, when I dabbled in crosses between merinos and coarse breeds, that a ram which was the produce of in-and-in breeding stamped his properties on the mongrel offspring with peculiar force; and I am not certain this rule does not obtain to some degree among full-bloods. I am inclined

4

to question whether the great cavanas of Spain, some of them once numbering 40,000 sheep, would ever have acquired their remarkable identity of characteristics without that in-and-in breeding to which they were subjected. Some intelligent observer of them in Spain, fifty or sixty years ago, whose name I do not now remember, said that in every hundred were were ten rather better and ten rather worse ones, but that the other eighty could hardly be distinguished one from another.

" The second property I have noticed in the ram, which gives him the power strongly to impress his qualities on his offspring, is constitutional vigour. He should be thoroughly masculine. He should be compact and massive in every part, his large scrotum almost sweeping the ground. He should not have a particle of a ' ewe look ' about him. Even his fleece should not be as fine as a ewe's fleece. He should have strength to knock down an ox. He should have undaunted courage, and delight in battle—fighting with desperate determination until slain or acknowledged master of the flock. Other things being equal, such are more usually, according to my experience, the rams which transmit their characteristics to their descendants.

" A ram of no extraordinary individual qualities is sometimes found to be a remarkable sire. He who obtains one of those highly valuable sires should cling to him as he would to gold, whether individually he ranks in the first or second class. This ' marking ' property is sometimes carried so far that a familiar and observing eye will promptly detect its effects in a strange flock, picking out every animal got by the particular ram, and even picking out his descendants, if bred among each other, for all subsequent generations."

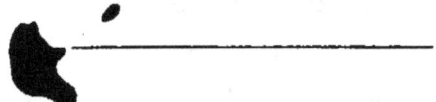

CHAPTER IV.

CROSSING.

Cross-Breeding between the Merino and Coarse Breeds.— On this subject we again quote from Mr. Randall :—

" The range of cross-breeding between fine and coarse-woolled sheep is comparatively limited, because there is but one breed of the former of any recognized importance, viz., the merino. And no intelligent man at the present day would any more think of crossing the merino with another breed to improve the characteristics sought in the merino than he would of alloying gold with copper to improve the qualities of the gold.

" When the object of such crossing has been to improve coarse, inferior races, it has succeeded for certain purposes. The coarse, common sheep of our country, for example, are always rendered more valuable by the infusion of merino blood. They gain materially in fleece, and lose in no other particular. But all crosses between merino and large, early maturing, improved English breeds and families, such as the Leicesters, Cotswolds, and the different families of Downs, have uniformly resulted in failure, and must always do so, as long as the characteristics of the respective breeds remain the same. The largest and heaviest-fleeced merinos would probably increase the weight of fleece of even the heaviest-fleeced English long-wools, but the wool loses by the cross its present specific adaptation to a demand always great in England and now rapidly increasing in the United States.

" The mutton is not injured—nay, for American tastes, it is decidedly improved by the cross ; but the long-woolled sheep loses its size, its early maturity, its propensity to fatten, and its great prolificacy in breeding. It loses the faultless form of the English sheep, without even acquiring the knotty compactness of the merino. In short, in the expressive common phrase, it becomes 'neither one thing nor the other,' but only a comparatively valueless mongrel between two—for their own separate objects—unimproved breeds !

" The cross between the merino and the Down materially increases and improves the fleece of the latter.   But it is held to detract from the value of the mutton, and it seriously impairs the value of long-wools.

" All attempts to establish *permanent intermediate varieties* of value by crossing between the merinos and any family of mutton sheep, with a view of combining the especial excellencies of each, have ended in utter failure.   Those with the Down and the Ryeland seemed to promise best, yet they not only resulted in disappointment, but produced mongrels incapable of being bred back to either of the English types.

" The merino, owing doubtless to its greater purity of blood compared with most other breeds, and to its vastly greater antiquity of blood compared with any of them, possesses a force and tenacity of hereditary transmission which renders it a most unmanageable material in any cross aiming at middle results. Its distinctive peculiarities are made to give way with difficulty, and its tendency to breed back is sometimes unconquerable. But if the merino fuses with reluctance, it absorbs other breeds with rapidity. A cross between it and a coarse breed is also legitimate and successful where the object is to merge that coarse breed entirely in the merino. This is accomplished by putting the ewes of such breed, and every new generation of their crossbred descendants, steadily to pure-blood merino rams.

"In such crosses the high qualities of choice rams render themselves eminently conspicuous—even more so, relatively, than in breeding among full-bloods. The descendants of such rams in the second cross ($\frac{3}{4}$ blood) are frequently more valuable than those of mediocre rams in the fourth or fifth cross ($1\frac{5}{16}$ or $3\frac{1}{32}$ blood).

"In the matter of profit these grades approach the full-blood rapidly. But there never was a more preposterous delusion than that entertained by the early French breeders, that 'a merino in the fourth generation ($1\frac{5}{16}$ blood), from even the worst woolled ewes, was *in every respect* equal to the stock of the sire.' Chancellor Livingston, who asserts this to have been the opinion of the French breeders, further says :—'No difference is now (1809) made in Europe in the choice of a ram, whether he is a full-blood or fifteen-sixteenths.' This undoubtedly solves problems in relation to a portion of the French merinos which otherwise would be quite inexplicable. They are, undoubtedly, *grade* sheep. The Germans, on the other hand, refuse to the highest-bred grade sheep any other designation than 'improved half-bloods.' 'They found,' says Mr. Fleichmann, ' that their original coarse sheep had 5,500 fibres of wool on a square inch of skin; that grades of the third or fourth merino cross have about 8,000; the twentieth cross, 27,000; the perfect pure-blood, from 40,000 to 48,000.' I do not apprehend that there is anything like an equal difference between the number of fibres on a given surface of the American merino and its grades; but in thirty years' observation of such grades, of every rank— some of them higher than the tenth cross, where there is but one part of the blood of the coarse sheep to 1,023 parts of merino blood—I never have yet seen one which, in every particular, equalled a full-blood of the highest class.

" Notwithstanding the brilliant and frequent successes in crossing different merino families (especially where the object is to merge an inferior into a superior family), the failures, or comparative failures, have been far more numerous. To cross different families of any breed merely for the sake of crossing, under the impression that it is *in itself* beneficial to health, or in any other particular, or with a vague hope that some improvement of a character which cannot be anticipated may result from it, is the height of folly and weakness. Even uniform mediocrity is far preferable to mediocrity without uniformity; and he who has the former should not break it up by crossing, without having a definite purpose, a definite plan for attaining that purpose, and enough knowledge and experience on the subject to afford a decent prospect of success. It is always safer and better, in seeking any improvement, to adhere strictly to the same breed and family, if that family contains within itself all the requisite elements of the desired improvement, or as good ones as can be found elsewhere. The most splendid successes among all classes of domestic animals have been won in this way. Successful crossing generally requires as much skill as successful in-and-in breeding. And as it is vastly more common, so vastly more flocks have been impaired in value by it, or at least hindered from making any important and permanent improvement. They are not permitted to become established in any improvement before it is upset by a new cross ; and these rapid crosses finally so destroy the family character of the flock—infuse into it so many family and individual strains of blood to be bred back to —that it sometimes becomes a mere medley, *which has lost the benefit that blood confers*—viz., family likeness, and the power to transmit family likeness to posterity.

"Every flockmaster or breeder should, after due observation and reflection, fix upon a standard for his flock—a standard of form, of size, of length of wool, of quality of wool, &c., &c.; and on this he should keep his eyes as steadily as the mariner keeps his eyes on the lighthouse in the darkness on a dangerous coast. Even in using a fresh ram from an unrelated flock of the same family (which is not crossing), he should use one which

conforms as nearly as possible to his standard. If he
disregards this; if he uses rams now tall and long-
bodied, and now low and short; now short and yolky-
woolled, and now long and dry woolled; now fine and
now coarse—in a word, each varying from its predecessor
in some essential quality—he will not, perhaps, break up
his flock quite as much as he would by *crossing* equally
at random, but he will do the next thing to it; he will
give it an unsettled and unhomogeneous character, and
materially retard, if not altogether prevent, essential
improvement.

" *Crossing between English Breeds and Families.*—
If we assume, with Mr. Youatt, that the long and
short woolled sheep of England are each respectively
descended from common ancestors, they form but two
breeds of sheep. There have been but very few successful crosses between these two breeds. The Hampshire
and Shropshire Downs, however, both ranked as first-
class sheep, and both officially classed as short-wools,
have usually a dip of long-wool blood. The Oxfordshire
Downs are the result of a direct cross between the Down
and the Cotswold, and they are already claimed to be an
'established variety.' But the instances of failure in
blending the breeds have been so much more numerous
than the successes that the balance of intelligent opinion
seems to be decidedly against such attempts. With
them, as with the merino, the successes in crossing
between the different families of the same breed have
been numerous and signal. Mr. Bakewell, there is little
doubt, was the first great improver in this direction,

though we are scarcely authorized to cite his examples, because, with a spirit much better befitting a nostrum vendor than a reputable breeder, he veiled all his proceedings in the closest mystery, and even permitted the knowledge to die with him. Some, therefore, have affected to believe that he resorted to different breeds, as he is known to have done to different families, in selecting his materials. But there are no proofs of the fact, and all the probabilities favour the conclusion that he adhered strictly to the long-woolled families. Among the facts which would seem by analogy to favour the latter conclusion was his own rigid in-and-in line of breeding after his materials were selected. If he deemed such quasi-identity both in blood and structure necessary or favourable to the completion of his object, it can scarcely be supposed that he would have voluntarily, and wholly *unnecessarily*, disregarded so great a discrepancy as that of a total difference in breed in its outset, or even that he would have spread his selection over any unnecessary number of families within the same breed.

"Mr. Bakewell's improved Leicesters have, since his death, again been improved by a dip of Cotswold blood. It is found to invigorate their constitutions, and render them better in the hindquarters. The Cotswolds of the present day have generally been rendered a little more disposed to take on fat rapidly, and to mature earlier, by a Leicester cross. The new Oxfordshire sheep is but a Cotswold improved by Leicester blood.

"The Hampshire and Shropshire Downs may be cited

as conspicuous examples of successful crossing between the short-woolled families, for it is mainly to those families that we owe their peculiar excellence, and not to any strain of long-wool blood where it exists in them. Various of the minor British short-woolled families have also been improved by crosses with the Down.

"For another and merely temporary purpose—viz., to obtain larger and earlier lambs or sheep for the butcher—it is legitimate to cross between different breeds or families indiscriminately, where the object in view can be effected in the first cross. The nature of the soil, food, or climate may be unfavourable to the large, early-maturing mutton families, but sufficiently favourable to some smaller and hardier sheep. Indeed, many such localities, in all old countries, have families, grown on them for many generations, which have gradually become so adapted to their surroundings that conditions highly unfavourable to other sheep have become innocuous, if not actually favourable, to them. Yet these local families may be ill adapted to meet the requisitions of the most accessible mutton markets, or, indeed, of any mutton market. They may be too small, too late in maturing, too indisposed to take on flesh, fat, &c., &c. In such cases rams of an improved mutton family—the family being selected with especial reference to the demands of the particular market and the defects to be counteracted in the local family—are put to the ewes of the local family, and the produce, as is usual with half-bloods, partakes strongly of the physical properties of the sire, and yet retains enough of the hardiness and local adaptation of the dam to thrive and mature where the full-blood or high-bred grade of the superior family could not do so. But in all such instances the grower should stop with the first cross. If, seduced by the beauty of that cross, he makes a second one between the full-blood and half-blood females, he obtains animals very little better than their dams for the purposes of mutton sheep, and decidedly less adapted to the local circumstances. Accordingly, some portions of the local family should also be bred pure by themselves to furnish females for the cross. This last course is generally pursued among the breeders of England who make such crosses.

" An analogous course of crossing might be resorted to with great profit by those farmers who *prefer* to make mutton production the leading object of their sheep husbandry, and who now grow those immense flocks of ' common sheep.' A single proper cross of English blood on these sheep would produce a stock which it would cost little more to raise than it now costs to raise common sheep *in the most profitable way*, and which would habitually command 50 per cent. more in market, and be ready for market a year earlier, than the common sheep. The value of the wool would not be lessened by any of the proper English crosses, and would be considerably increased by some of them.

" The selection of the English family for the purposes of the above cross should be made with strict reference to local circumstances. On rich, sufficiently moist lands, not subject to summer drought, bearing an abundance of the domesticated grasses, and near good local mutton markets, the unrivalled earliness of maturity in the Leicester would give it great advantages; but it would bear no even partial deprivation of feed, no hardships of any kind, and no long drives to distant markets. The Cotswold is a hardier, better working and driving sheep, inferior to the Leicester in no particular which would be very essential in such situations; and I cannot but think that, for the object under consideration, those sub-families of it which have not been too deeply infused with Leicester blood offer essential materials for a cross. The different Down families will bear shorter keep than the preceding, and will range over larger surfaces to obtain it. They are considerably hardier than the Leicesters, or those families of the improved Cotswolds which have much Leicester blood. They can endure slight and temporary deprivation of food better than the long-wools; but it is a mistake to suppose that any mutton breed or family will fully or profitably attain the objects of its production without abundance of suitable food being the rule, and deprivations of it any more than the occasional exception. (I speak, of course, of sheep which are grown only for the butcher, the leading objects of whose production is high condition and early maturity.) The Downs also produce better mutton; and the dark legs and faces of the half-bloods always give them a readier and better market. But the half-blood Downs would generally carry less wool than the half-blood long-wools.

" In hardiness, patience of short keep, and adaptability to driving long distances, any of the half-bloods would surpass their English ancestors, and would, under the conditions already stated, generally flourish vigorously.

"Though the crossing of mutton breeds has, in many instances, entirely different objects from those sought in crossing sheep kept specially for the production of wool, and though, consequently, the proper modes of crossing in the two cases often vary essentially, still the general views expressed in regard to unmeaning, aimless, and unnecessary crossing are as applicable to the English mutton sheep as to the merino."

CHAPTER V.

SELECTION OF SHEEP.

For the following we are again indebted to Mr. Randall :—

"CARCASS.—Carcass is unquestionably the first point to be regarded, even in fine-woolled sheep—because on the proper constitution, or the proper structure and connection of its parts, depends the health, vigour, and hardiness of the animal; and without these all other qualities are as houses built on sand. Plump medium size, for the breed or variety, is the most desirable one. The body should be round and deep, not over long, and both the head and neck short and thick. The back should be straight and broad; the bosom and buttock full; the legs decidedly short, well apart, straight, and strong, with heavy fore-arm and fulness in the twist.

This 'pony-built' figure indicates hardiness, easiness of keep, and a predisposition to take on flesh.

"SKIN.—The skin should be of a rich, deep rosy colour. The Spaniards ever justly regarded this a point of much importance, as indicative of fattening or easy-keeping properties of the animal, and of a normal and healthy condition of the system. The skin should be thinnish, mellow, elastic, and particularly loose on the carcass. A white skin, when the animal is in health, or a tawny one, is rarely found on a high-bred merino. A thick, stiff, inelastic skin, like that found on many under-bred French sheep, is highly objectionable.

"FOLDS.—The Spanish, French, and German breeders approve of folds in the skin, considering them indications of a heavy fleece. The French have bred them over the entire bodies of many of their sheep. I have seen two hours and a half expended by an active and skilful shearer in getting the fleece decently off a ram of this stamp. A deep, soft, plaited dewlap on both sexes, and some slight comigation on the neck of the ram, were all our old breeders of the merino desired in this way. The fashion has extended to heavy neck folds, particularly on the ram, a short fold or two on the back of the elbow, and some small ones round and on the roots of the tail, and on the breech—the latter running in the direction of lines drawn from the tail to the stifle. Gentle corrugations over the body, which can be pulled smooth in shearing, are also tolerated.

"FLEECE.—Wool long enough to do up in the fleece is not desirable on the nose, under the eyes, or on the legs below the knees and hocks, though a thick coat of shortish wool on the latter, and particularly on the hind legs, is regarded as a good point. The arm-pits and most of the base of the scrotum must necessarily be bare; but these cavities should be as small as the freedom of movement permits; and all the other parts of the body and limbs should be densely covered with wool of as uniform a length as is attainable. It is a specially fine characteristic to see it of full length on the belly, forehead, and cheeks, and on the legs as far down as the knees and hocks.

"The wool should stand at right angles to the surface, except on the inside of the legs and on the scrotum; it should present a dense, smooth, even surface externally, dropping apart nowhere; and the masses of wool between those natural cracks or divisions which are always seen on the surface should be of medium character. If they are too small, they indicate a fineness of fleece which is incompatible with its proper weight; if too large, they indicate coarse, harsh wool.

"The good properties of wool are too well understood to require many words. Length is no longer an objection to the finest staple, as it once was. The maximum, both of thickness and length, cannot be attained on the same animal, and the object of the breeder should be to produce that particular combination or co-existence of these properties which will give the heaviest fleece.

"FINENESS.—The grower knows his market, and must produce an article adapted to it. In the American market there is a much larger demand for medium than

fine wools, and the former commands much the best
price in proportion to cost of production. It is to be
hoped, however, that the demand for fine wools will
increase. Whatever the quality aimed at, it should be
the same throughout the flock, so far as it is practicable.

"EVENNESS.—Evenness of quality in every part of
the fleece, so far as this can be attained, is one of the
first points of a well-bred sheep. Tar is very objectionable, but not so much so as what the Germans term dog's
hair—hair growing out through the wool on the thighs,
the edges of the neck folds, about the roots of the horns
in rams, or standing scattered here and there through
the fleece, or inside the legs. This indicates bad blood,
or a defective course of breeding.

"FINENESS AND SOUNDNESS.—Wool should be of equal
diameter from the root to the point of the fibre. It
should be especially free from any finer or weaker spot,
or 'joint,' in it, occasioned by a temporary illness, or
other low state of the animal. This can often be
detected by the naked eye, and always by pulling the
fibre. Wool is said to be sound when it is strong and
elastic.

"PLIANCY AND SOFTNESS are considerations of the
first importance, not only as indicia of other qualities,
but intrinsically. If we can suppose two lots of wool
exactly to resemble each other in every other particular,
but that under the same treatment one is comparatively
stiff and hard to the touch, while the other has a silky
pliancy and softness, the latter is decidedly the most
valuable, because it will produce manufactured articles

far superior in beauty and for actual use. But, in point of fact, full-blood wool is invariably soft in proportion to its fineness, and is always so in proportion to its marketable value. A practised buyer can decide on that value in the dark.

"STYLE is, perhaps, a word which has rather vague boundaries to its meaning; but it includes that combination of useful and showy properties which give value to the choicest wool—viz., fineness, cleanness of colour, lustre, uniformity and beauty of covering, and that peculiar mode of opening on the body, or disposition of the fibres in the shorn fleece, which indicate the last extreme of pliancy and softness. These qualities, in combination, present an appearance which at once, without a sufficiently close inspection to discover the separate fibres, or even without a touch of the hand, point out the best fleece in the pile.

"YOLK.—Vauquelin, a celebrated French chemist, found that various specimens of yolk contained about the same constituents :—1. A soapy matter with a basis of potash, which formed a greater part of it. 2. A small quantity of carbonate of potash. 3. A perceptible quantity of acetate of potash. 4. Lime, whose state of combination he was unacquainted with. 5. An atom of muriate of potash. 6. An animal oil, to which he attributed the peculiar odour of yolk. He found the yolk of French and Spanish merinos essentially the same. He assumed that the yolk in shorn wool injures it after a few months, if not scoured out.

"Different opinions are entertained of the amount of

yolk it is profitable to propagate in wool. If the fleece is sold unwashed, and, according to the present general mode, at a fixed rate of shrinkage on that account, it is obviously the interest of the wool-grower to produce as much yolk as is consistent with the greatest united production of wool and yolk. And even if wool is sold nominally 'washed,' it is evident that the same amount of washing will leave very yolky fleeces heavier than unyolky ones. Farmers have learned that, if they can only *say* their wool is washed—no matter *how* washed—10 or 15 per cent. more yolk than would be left by the thorough washing will not cause any corresponding deduction in the price. There are a class of experienced buyers, certainly, who do not purchase in this indiscriminate way; but as the wool business has constantly expanded, and opened new opportunities for the profitable investment of money, every year brings its fresh horde of raw, eager buyers—the agents of manufacturers or speculators, or persons speculating on their own account—and some of these always take the heavy, dirty wools at about the price of the clean ones.

" It is particularly fortunate for the preservation of the intrinsic value of our merino sheep, and fortunate for the public interest, that it is already incontestibly ascertained that the greatest amount of yolk is not consistent either with the greatest amount of wool, or with the greatest aggregate amount of both yolk and wool. The black, miserably 'oily,' 'gummy' sheep, looking as if their wool had been soaked to saturation in half-inspissated oil, and then daubed over externally with a coating of tin and lamp-black, never exhibit that maximum of both length and density of wool which, with a proper degree of yolk, produces the greatest aggregate weight.

" Yolk has been generally thought to be the pabulum of wool; and if so, its excessive secretions, as a separate substance, may diminish its secretions in the form of wool. Be this as it may, the fact I have stated stands without an exception. And animals exhibiting this marked excess of yolk are invariably feebler in constitution, less easily kept, and especially less capable of withstanding severe cold. Such excessive secretions appear, then, to cause, or else to be the result of, an abnormal or defective organization. For these reasons these comparatively worthless animals, once so eagerly sought, have already gone out of use among the best-informed breeders; and where they linger, it is, like antiquated fashions, in regions where the current ideas of the day penetrate slowly!

" There should be enough fluid yolk within the wool on the upper surfaces of the body to cover every fibre like a brilliant, and, in warm weather, like an undried coat of varnish, but not enough to fill the interstices between them; so that the fleece shall appear, as it sometimes does, to be growing up through a bed of oil. And if there is a sufficiency of yolk above, it must be expected that, underneath, where the fleece is less exposed to evaporation and the washing of rains, and to which part gravitation would naturally determine a fluid substance, a considerably greater quantity of it will be found; but hardened or pasty masses of it within the wool are to be avoided on all parts of the body. A portion of the fluid yolk will necessarily inspissate, or harden, on the outer ends of the wool. It is proper that it should sensibly thicken those ends, and clot them together in small masses on the upper parts of the body, forming a coat considerably thicker, firmer, and harder to the hand than would be the naked wool, and quite rigid when exposed to cold ; but it should not cover the wool in round knobs, or in thick, firmly-adhering patches, bounded by the fleece cracks—sticking to the hand, in hot weather, like a compound of grease and tar, and in cold having a ' board-like' stiffness. Underneath, for the same reasons given in reference to inside yolk, a greater quantity of it must be tolerated. It should stick the masses of wool together in front of the brisket and scrotum ; and large rounded knobs of it inside the legs and thighs, and on the back side of the scrotum, are considered desirable.

" COLOUR OF YOLK.—The external yolk is occasionally somewhat yellowish—of the tinge of dirty bees-wax— but more generally of some dark shade of brown, or what would more commonly be termed black. The darker colour is preferred. All American merino sheep having what is esteemed a sufficient amount of yolk, become very dark coloured each year before the winter is far advanced if they are housed from summer and winter storms after shearing. Rains wash away the yolk, and with it the colour. But the yolk is soluble in different degrees in different families, and even on different animals of the same flock. The Paular (Rich) sheep hold their colour uncommonly well ; the French rapidly bleach. It has been supposed that the black colour is communicated to external yolk by dust. This may contribute to the result, but the change in colour is partly chemical.

" Internal yolk varies in colour from a pure white to a pure yellow. It has been rather the fashion, since the days of the Saxon sheep, to breed for the former, and this is the prevailing colour in the American Paulars. The breeders of the American Infantados, and of the Silesians, generally follow the old Spanish custom of giving preference to shades of yellow. A brilliant ' golden tinge,' faint or imperceptible near the roots of the wool, but deepening towards its outer extremities, is the one sought after. The founder of the improved Infantado family has bred steadily for that colour ; and he has done so not merely as a matter of taste, but under the impression that it betokens a vigorous growth of wool and a vigorous constitution, and particularly vigour of that kind which exhibits itself in the forcible trans- mission of individual properties to progeny. But this ' golden tinge ' is not to be mistaken for the deep saffron yellow which attends cotting, or for a dull dead yellow, or for a tawny bees-wax hue, or for the hue of ' nankeen cloth,' sometimes seen in imperfectly-bred animals. The favourite colour among the French breeders is a creamy one. In answer to inquiries made by me, several experi- enced manufacturers—all I consulted—concurred in the statement that the colour of the yolk is not in itself a matter of any consequence, in reference to any of the objects of manufacturing ; and that its quantity and

consistency are only important in so far as they affect
its weight, and cause a loss in scouring.

"In many regions where sheep are not pastured on
thoroughly soddened ground, the whole interior of the
fleece becomes stained by dust to the prevailing colour
of the ground."

CHAPTER VI.

WOOL.

WOOL is a variety of hair. The term hair is applied,
in ordinary language, to smooth, straight-surfaced
filament, like human or horse hair, with serrations of
any kind on its surface. Wool, on the other hand, is
always more or less waved, besides which, externally,
each wool and filament is seen, under the microscope, to
be covered with scales, overlying each other, and projecting wherever a bend occurs in the fibre. Upon the
minute points of difference here shown the value of
wool chiefly depends, especially with regard to the great
variety of its applications.

If each fibre were straight and smooth, as in the case
of hair, it would not attain the twisted state given to it
by spinning, but would rapidly untwist when relieved of
the force used in spinning ; but the waving condition
causes the fibres to become entangled with each other,
and the little projecting points of the scales hook into
each other, and hold the fibres in close contact. Moreover, the deeper these scales fit into one another, the

closer becomes the structure of the thread, and, consequently, of the cloth made of it.

This gives to wool the quality of *felting*. By combing, or drawing the wool through combs with angular metal teeth, some of the scales are removed, and the points of many more are broken off; so that wool which has been combed has less of the felting property, and is consequently better adapted for the manufacture of light fabrics. The yarn made of such wool is called *worsted*, and the cloths made of it are called *worsted goods*.

But such is the variety of wool attained by careful breeding and selection, that these differences can be got without combing — some wools being found to have naturally fewer serratures, and a less wavy structure, than others.

These are consequently kept separate, and are called *combing-wools;* whilst those which are much waved, and have many serratures, are called *carding-wools*, from their being simply prepared for spinning by a carding-machine.

The serratures, or points of the scales, are exceedingly small, and require the aid of a good microscope to see them. They vary from 1,200 up to 3,000 to an inch.

Wool is the most important of all animal substances used in manufactures, and ranks next to cotton as a raw material for textile fabrics. Its use as a substance for clothing is almost universal in the temperate regions of the globe.

Previous to 1791, British woollen cloths were made

almost wholly of native-grown wools. At that time the whole supply of the country could not have much exceeded 100,000,000 lbs. The merino wool of Spain then began to displace them in the best kinds of goods, and the imports from that country reached their maximum in 1805, being in that year 7,000,000 lbs.

Before 1820 the German wool began to supersede the Spanish, and was imported largely until 1841. After that, the cheaper wool of the British colonies, to a great extent, took the place of the German; and the latter is now chiefly used for only the finest cloths.

Wool varies in character according to the peculiar breed of sheep which yields it, and also with the nature of the soil, food, shelter, and climate. In a wool of first-class quality the fibres are fine, soft, elastic, sound, of good colour, and free from deleterious or troublesome impurities. The commercial value of any sample depends, therefore, upon the extent to which it possesses these qualities. If it be a combing-wool, it will also depend upon its length of staple.

For technical purposes, shorn fleeces are divided into two classes—one called " hogs," or " tegs ;" the other, " wethers," or " ewes." The former are the first fleeces shorn from the sheep; the latter are those of the second and succeeding years. But the meaning of these terms varies in different districts. The fleeces of yearlings are, as a rule, longer in the staple, and otherwise of superior quality to the wool of older animals.

Wool taken from the skins of slaughtered sheep is called *skins-wool* or *pelt-wool*, and is of more variable quality than fleece-wool, on account of its being obtained in all stages of growth.

As long-stapled wools are used for worsted goods, and short-stapled for woollen goods, the various breeds of sheep which yield these two leading kinds are naturally divided into the long-woolled and short-woolled classes of sheep.

The Lincoln, the Leicester, and the Cotswold breeds are considered good types of the former, and the Downs, the Welsh, and the Shetland breeds of the latter.

The following brief notice of the characteristic properties of the various wools native to Great Britain is founded upon the description given of them in the Jury report of the International Exhibition of 1862, held in London.

Of the " long wools," the Lincoln has greatly risen in value of late years. It is coarse, of great length, and silky in appearance, so that it is well adapted for lustre goods, in imitation of alpaca fabrics. The Leicester wool is highly esteemed for combing. It is rather fine in the hair, but not usually so soft and silky in the staple as the last. Cotswold wool is similar to the Leicester, but somewhat harsher. It is not suited for lustre goods. Highland wool is long stapled and of coarse quality, but known to be susceptible to great improvement. The practice of smearing greatly depreciates its value. It is chiefly used for the coarsest kinds of woollen fabrics, as carpets, rugs, and similar articles. It is also used for Scotch blankets.

Of the " short wools," the different breeds of Down

sheep partake very much of the same characteristics, but soil and climate affect them. The *Southdown* is a short-stapled, small-haired wool, the longer qualities of which are put aside for combing purposes, and the shorter for the manufacture of light woollen goods, such as flannel. The *Hampshire Down* differs from it in being coarser, and in having the staple usually longer. The *Oxford Down*, again, exceeds the last in length and coarseness of staple. The *Norfolk Down*, on the other hand, when clean, is of a very fine and valuable character. The *Shropshire Down* is a breed increasing in importance, and the wool is longer in the staple, and has more lustre, than any other of the Down breeds. *Ryelands* wool is fine and short, but the breed is now nearly extinct. The *Welsh* and *Shetland* wools have a hair-like texture, deficient in the spiral form, upon which depends the relative values of high-class wools. They are only suited for goods where the properties of shrinking and felting are not required. Shetland wool is obtained of various natural tints, which enables it to be used for producing different patterns without dyeing. Of the intermediate wools, *Dorset* is clean, soft, and rather longer, and not quite so fine in the staple as the Down breeds. The *Cheviot* has increased very much of late years in public estimation. It is a small, fine-haired wool, of medium length, and is suitable for woollen and worsted purposes, for which it is largely used.

Some of the British colonies are very important wool-producing countries, Australia, in this respect, standing far in advance of all other countries whatever. The

Australian wool has, in general, a beautiful, short, silky staple, well adapted for the manufacture of soft, pliable, and elastic fabrics. All the settled districts of the island-continent have been found well suited to the growth of fine-woolled sheep, and the extraordinary increase of the flocks forms one of the most remarkable features of the country.

The breed has sprung from one merino ram and five ewes, brought out, as before stated, by Mr. M'Arthur, in 1796.

The alpaca wool grown in Australia since the creature was first introduced, some years ago, is of inferior quality; but this is said to have arisen from rearing the animals too near the sea-coast, and hopes are entertained of succeeding with it better inland.

The wool of Cape Colony has of late years been much improved by the introduction of merinos, and, as will be seen from the table below, the exports from it are increasing very rapidly.

Among the exports of Great Britain from India, wool has of late become an important article, the quantity having risen from 2,500,000 lbs. in 1840 to nearly 19,000,000 lbs. in 1874, but the supply is rather fluctuating. A great deal of the Indian wool is coarse and hairy, and can only be used for low-class goods. We may state here that the most costly of all wools is obtained from the Thibetan goat, and is found next the skin, under the thick hair of the animal. From it the far-famed Cashmere shawls are made. The highest price of any quality which is sold is from 6s. to 7s. per lb. in the native markets, but the Maharajah of Cashmere keeps a strict monopoly over the best kinds.

Turning now to European countries, it is somewhat sad to think that Spain, the native country of the merino, which not so long ago sent all the wool for the finest English cloths, has allowed its quality to degenerate, and its large supply to dwindle away.

The wool of Saxony, Silesia, and some parts of Austria, which is obtained from sheep originally of the pure Spanish merino breed, is the finest produced in any country; and, notwithstanding the lower price and nearly equal quality of the Australian, German wool is still employed for the manufacture of the finest broadcloths, some kinds of ladies' shawls, and a few other purposes.

Great attention is paid to the breeding and rearing of sheep in Germany, and large flocks are reared for their wool alone. In Austria the number of sheep is estimated at 45,000,000, and the annual yield of wool at 100,000,000 lbs., most of it being of fine quality, and all of which is consumed in Austrian manufactures. France produces a large quantity both of fine and of coarse wools.

In Italy the production of wool from mixed merino breeds has become a source of great wealth. Russia, as might be expected from its great extent, rears many qualities, from the finest merino to a very coarse kind. The wools of the remaining countries of Europe are of minor importance.

We must not omit to mention that the wools of South America are now attaining great importance, as will be seen from the table below; but it is necessary to state that, besides the 10,562,874 lbs. imported in 1874, there were 4,182,867 lbs. of alpaca (including llama and vicuna) wool. The wool of the alpaca is very fine, from 6 to 12 inches long, of various colours, and well suited for various kinds of goods. South American sheep's-wool is of an inferior quality.

Much finer wool would be produced in Great Britain than is at present if it were not that the demand for mutton and the unfitness of the merino sheep for supplying that article of good quality, lead the farmers of the home country to choose breeds which are primarily mutton-producers.

The following table will show at a glance the remarkable changes which have taken place in the sources from which Great Britain has derived its supplies of wool, and also the steady increase in the aggregate quantity imported into that country. We regret our inability to supply data of a similar nature up to a later date.

IMPORTS OF WOOL TO GREAT BRITAIN FROM THE PRINCIPAL COUNTRIES.

Year.	Spain.	Germany.	Australia.	S. Africa.	E. Indies.	S. America.
	lbs.	lbs.	lbs.	lbs.	lbs.	lbs.
1810	5,952,407	778,853	167	—	—	—
1816	2,958,607	2,816,655	13,611	9,623	—	—
1820	3,536,229	5,113,442	99,415	29,717	—	—
1830	1,643,515	26,073,882	1,967,279	33,407	—	—
1834	2,343,915	22,634,615	3,558,091	141,707	67,763	—
1840	1,266,905	21,812,099	9,721,243	751,741	2,441,370	4,378,274
1850	440,751	9,166,731	39,018,221	5,709,529	3,473,252	5,296,648
1860	1,000,000	9,292,000	59,166,000	16,574,000	20,214,000	8,950,000
1870	25,262	4,912,600	175,081,427	32,785,271	11,143,148	12,475,631
1874	100,178	7,158,319	225,383,631	42,232,672	19,116,772	10,562,874

To get the total imports for each year, we would
require to add the amounts from countries of lesser
importance, which we cannot do, not having the requisite
statistical information; but in the next statement we
give the total annual imports for the three years 1872,
1873, 1874. They were—in 1872, 302,500,925 lbs.; in
1873, 313,496,742 lbs.; and in 1874, 340,288,032 lbs.

For several years past about one-third of the British
wool imports have been re-exported. The estimated
produce of home-grown wool, in 1871 and the three preceding years, was as follows :—

In 1871. Total number of animals
    slaughtered   12,370,056
Estimated at 2¾ lbs.   ...   .   34,017,654 lbs.
1871. Net clip of wool...   ...   ...   144,985,712 ,,
1870.   ,,   ,,   ,, ...   149,516,679 ,,
1869.   ,,   ,,   ,, ...   155,591,096 ,,
1868.   ,,   ,,   ,, ...   165,549,735 ,,

Thus showing a gradual annual decrease in the net clip
of wool. It is to beregretted that further information
regarding the numbers of sheep slaughtered, and the
quantity of wool thereby obtained, is not at hand.

Independent of the vast quantity of home and foreign
grown wool which finds its way to the markets of the
universe, as wool that is in fit condition for spinning
and weaving, considerable quantities are retained on the
skins, and made into rugs and mats for house and
carriage use. For this purpose skins of the very best
quality are chosen, and it is necessary that the wool
should be long in the staple. After being carefully

curried, the long silky locks of wool are dyed, usually some bright colour, and combed. The skins are pared to shape, and form handsome rugs, which are not only in great favour in Great Britain, but are extensively imported. The chief seat of this trade is at Bermondsey, in London, but it is also carried on to a considerable extent in other parts of the kingdom. Large numbers of Australian sheep and lamb skins, usually black, are also imported, in the wool, and are dressed and used as furs, that is, for personal wear; and some of the lambs' skins for this purpose fetch high prices.

With respect to the wool, or woolly hair of animals, other than the sheep, which we have not already mentioned, the only one of much importance is mohair, or the wool of the Angora goat. Of this material there were about 7,000,000 lbs. imported into Great Britain in 1874. It is a white silky wool, with an average length of staple of from 5 inches to 6 inches. The demand for it is only of recent origin, and it is chiefly used for certain kinds of ladies' dresses. The hair of camels, bullocks, common goats, and several furs, are also used for various manufacturing purposes to a considerable extent.

The grand total of wool, shoddy, and goats' hair employed in the woollen industries of Great Britain in 1874 cannot have been far short of 500,000,000 lbs. The total import of raw cotton in the same year was nearly 783,432,216 lbs.; but of this more than 129,483,816 lbs. were re-exported.

Proportion of Wool to Meat in Sheep of different Ages, Sexes, and Sizes.—The following paper we reproduce

from an American publication. It is from the pen of
Mr. Horner D. L. Sweet, a gentleman who has devoted
much time and care to the breeding of sheep, and who,
in the following tables, gives us the information derived
from a series of highly interesting and careful experiments.

"COMPARATIVE WEIGHT OF WOOL AND BODIES OF SHEEP.

" BY H. D. L. SWEET.

" The Hon. Robert R. Livingstone, the first president
of the first agricultural society of the State of New
York, in his justly celebrated essay on Fine-Woolled
Sheep, uses the following language :—'The inferiority in
the size of the merino to some other breeds, which some
make as an objection, is, in my opinion, an important
advantage, not only in sheep, but in every other stock
not designed for the draft ; because they will fatten in
paddocks in which larger cattle would suffer from the
fatigue they must undergo in order to procure the food
that is necessary for their support.

" ' This meaning applies more strongly to sheep than to
any other stock. They are generally kept upon high
and dry pastures, that are frequently parched in summer,
when fatigue is most irksome to them. To which we
may add that the fleece is not proportioned, as the food is,
to the *bulk* of the animal, but to his *surface*, and a small
sheep, having more surface in proportion to his bulk,
must also have wool in the same proportion. That is—
a sheep whose live weight shall be 60 lbs., and who, of

course, will require but one quarter of the food of a sheep that weighs 240 lbs., will, notwithstanding, have half as much wool (if the fleeces are equally thick) as his gigantic brother.'

"In proof of the first proposition—that sheep do consume in proportion to their bulk—Mr. Livingstone submits, in an appendix to his essay, the record of many experiments which show conclusively that such is the fact; but of his second proposition—that they shear in proportion to their surface, he gives no facts, and I presume it to be mere theory. The attention of the writer was called to this subject by the Hon. George Geddes, some few years since, and at his request the trial was made, and the result has been given to the world. Experiments of the same character on the same flock have been conducted for three successive years, and their results are recorded in the following tables.

"In one or two points they are not so perfect as I could wish, but they are the best that could be done with so small a flock. Had there been from forty to fifty in each class, and every year, the natural law in relation to them might be nearer in accordance with the facts noted; for, as there are exceptions to all rules, I may be giving the exception and not the rule. This can be true only in regard to five and six-year-old ewes and five-year-old wethers. In all other cases, taking the three years collectively, I am confident that facts of value have been obtained."

The tables already alluded to are given in the following pages.

SWEET BROS.' FLOCK.

TABLE I.—CLASSIFIED BY AGE AND SEX.

No. in Class.	Ages.	Sexes.		Gross Weight.	Weight of Bodies.	Weight of Wool.	Aver. of Bodies.	Aver. of Fleeces.	Pounds of Body to 1 of Wool.	Per cent. of Wool to Gross Weight.
		Ewes.	Wethers.							
19	1	E.	—	1,193·72	1,097	96·72	52·47	5·09	10·44	8·10
13	1	—	W.	965·23	894	71·23	68·77	5·48	12·55	7·37
15	2	E.	—	1,124·37	1,048	76·37	69·86	5·09	13·72	6·88
15	2	—	W.	1,383·92	1,299	84·92	86·66	5·66	15·29	6·53
9	3	E.	—	759·14	710	49·14	78·88	5·45	14·45	6·46
42	3	—	W.	4,155·11	3,891	264·11	92·64	6·28	14·73	6·88
41	4	E.	—	3,738·38	3,557	181	86·75	4·41	19·65	4·84
26	4	—	W.	2,921·13	2,736	185·13	105·11	7·12	14·76	6·33
180	1 to 4	84	96	16,341	15,331	1,010	85·17	5·83	15·17	6·18

TABLE II.—CLASSIFIED BY AGE AND SEX.

No.	Age.	Sexes.		Gross Weight.	Weight of Bodies.	Weight of Wool.	Aver. of Bodies.	Aver. of Fleeces.	Pounds of Body to 1 of Wool.	Per cent. of Wool to Gross Weight.
		E.	W.							
42	1	E.	—	2,378·57	2,189	189·57	52·11	4·51	11·60	7·96
52	1	—	W.	3,224·51	2,985	239·51	57·40	4·60	12·46	7·42
19	2	E.	—	1,387·16	1,292	95·16	68	5	13·57	6·86
13	2	—	W.	1,225·16	1,147	78·16	88·23	6	14·66	6·46
14	3	E.	—	1,026·31	960	66·31	68·57	4·70	14·47	6·46
13	3	—	W.	1,297·36	1,215	82·36	93·40	6·33	14·75	6·35
9	4	E.	—	726·59	679	47·59	77·44	5·28	14·26	6·54
27	4	—	W.	2,693·06	2,505	188·06	92·77	6·96	13·32	6·98
15	5	E.	—	1,178·15	1,111	67·15	74	4·47	16·54	5·77
11	5	—	W.	1,153·40	1,075	78·40	97·72	7·12	13·71	7·00
215	1 to 5	99	116	16,290·27	15,158	1,132·27	70·50	5·26	13·30	6·95

TABLE III.—CLASSIFIED BY AGE AND SEX.

No.	Age.	Sexes.		Gross Weight.	Weight of Bodies.	Weight of Wool.	Aver. of Bodies.	Aver. of Fleeces.	Pounds of Body to 1 of Wool.	Per cent. of Wool to Gross Weight.
		E.	W.							
14	1	E.	—	955·78	877	78·77	62·64	5·62	11·00	8·24
78	1	—	W.	5,623·84	5,201	422·84	66·67	5·42	12·30	7·71
42	2	E.	—	2,861·64	2,662	199·64	63·38	4·75	13·33	6·97
48	2	—	W.	3,994·79	3,735	259·79	77·81	5·41	14·37	6·50
33	3	E.	—	2,837·24	2,658	179·24	80·54	5·40	14·82	6·31
13	3	—	W.	1,338·89	1,251	87·89	96·23	6·76	14·23	6·56
13	4	E.	—	1,154·68	1,083	71·68	83·30	5·51	15·10	6·26
9	5	E.	—	735·93	680	45·93	75·35	5·10	14·82	6·24
10	6	E.	—	837·84	790	47·84	79·00	4·78	16·49	5·70
260	1 to 6	121	139	20,340·63	18,957	1,393·63	72·91	5·32	13·58	6·84

TABLE IV.—AVERAGE OF THESE THREE YEARS.

Classified by Age and Sex, the footing being the three flocks collectively.

No. in Class.	Age.	Sexes.		Average Weight of Body.	Average Weight of Fleece.	Pounds of Body to 1 of Wool.	Average Percentage.
75	1	E.	—	55·74	5·07	11·01	8·10
76	2	E.	—	67·08	4·94	13·54	6·90
56	3	E.	—	75·99	5·18	14·58	6·41
63	4	E.	—	82·49	5·06	16·33	5·88
24	5	E.	—	74·67	4·75	15·68	6·00
10	6	E.	—	79·00	4·78	16·49	5·70
143	1	—	W.	64·28	5·16	12·43	7·50
76	2	—	W.	84·23	5·69	14·77	6·49
68	3	—	W.	88·86	6·45	14·57	6·58
53	4	—	W.	103·94	7·04	14·04	6·65
11	5	—	W.	97·72	7·12	13·71	7·00
665	1 to 6	304	351	79·52	5·32	14·01	6·65

TABLE V.—CLASSIFIED BY WEIGHT.

In divisions of 10 lbs. each, except those weighing less than 50 lbs. and more than 100 lbs.

No.	Weight of Divisions.	Sexes.		Gross Weight.	Weight of Bodies.	Weight of Wool.	Aver. of Bodies.	Aver. of Fleeces.	Pounds of Body to 1 of Wool.	Per cent. of Wool to Gross Weight.
		Ewes.	Wethers.							
5	42 to 51	5	—	256	234	22	46·80	4·40	10·63	8·59
14	50 to 61	10	4	871	803	68	57·35	4·85	11·80	7·80
20	60 to 71	14	6	1,427	1,320	107	66	5·35	12·33	7·49
34	70 to 81	21	13	2,742	2,567	175	75·50	5·14	14·66	6·88
39	80 to 91	19	20	3,566	3,355	211	86	5·41	15·87	5·90
34	90 to 101	11	23	3,453	3,252	201	95·64	5·91	15·42	5·82
34	100 to 134	4	30	4,026	3,800	226	111·76	6·67	16·80	5·61
180	42 to 134	84	96	16,341	15,331	1,010	85·17	5·38	15·17	6·18

TABLE VI.—CLASSIFIED BY WEIGHT (as before).

No.	Weight of Divisions.	Sexes.		Gross Weight.	Weight of Bodies.	Weight of Wool.	Aver. of Bodies.	Aver. of Fleeces.	Pounds of Body to 1 of Wool.	Per cent. of Wool to Gross Weight.
		Ewes.	Wethers.							
37	34 to 51	23	14	1,875	1,725	150	46·60	4·05	11·50	8·00
41	50 to 61	19	22	2,460	2,270	190	55·37	4·63	11·94	7·72
42	60 to 71	25	17	2,940	2,740	200	65·23	4·75	13·70	6·80
30	70 to 81	24	16	2,432	2,272	160	75·73	5·33	14·20	6·57
25	80 to 91	6	19	2,266	2,110	156	84·40	6·24	13·52	6·88
25	90 to 101	2	23	2,568	2,408	160	96·32	6·40	15·05	5·84
15	100 to 127	—	15	1,743·27	1,633	110·27	108·86	7·35	14·80	6·32
215	34 to 127	99	116	16,290·27	15,158	1,126·27	70·50	5·26	13·30	6·95

TABLE VII.—CLASSIFIED BY WEIGHT (as before).

No.	Weight of Divisions.	Sexes.		Gross Weight.	Weight of Bodies.	Weight of Wool.	Aver. of Bodies.	Aver. of Fleeces.	Pounds of Body to 1 of Wool.	Per cent. of Wool to Gross Weight.
		Ewes.	Wethers.							
10	36 to 51	5	5	493	455	38	40·50	3·80	11·97	7·91
34	50 to 61	15	19	2,009	1,850	159	54·44	4·67	14·15	7·90
67	60 to 71	33	34	4,828	4,480	348	66·88	5·19	12·87	7·20
96	70 to 81	44	52	7,755	7,230	525	75·30	5·46	13·77	6·76
28	80 to 91	14	14	2,550	2,390	160	85·35	5·71	14·93	6·23
16	90 to 101	7	9	1,628	1,532	96	95·75	6·00	15·85	5·89
9	100 to 140	3	6	1,087·63	1,020	67·63	113·33	7·51	15·09	6·21
260	36 to 140	121	139	20,350·63	18,957	1,393·63	72·91	5·32	13·58	6·84

TABLE VIII.—THE AVERAGE OF TABLES V., VI., AND VII.

No.	Weight of Divisions.	Sexes.		Average Weight of Bodies.	Average Weight of Fleeces.	Pounds of Body to 1 of Wool.	Per cent. of Wool.
		Ewes.	Wethers.				
52	34 to 51	33	19	44·63	4·08	11·36	8·16
89	50 to 61	44	45	55·78	4·71	11·90	7·80
129	60 to 71	72	57	66·08	5·09	12·96	7·13
160	70 to 81	89	71	75·52	5·31	14·21	6·53
92	80 to 91	39	53	85·25	5·78	14·77	6·33
75	90 to 101	20	55	95·90	6·10	15·44	5·85
58	100 to 140	7	51	111·31	7·17	15·56	6·04
655	34 to 140	304	351	79·52	5·32	14·01	6·65

" The value of these tables can only be known by careful comparison, and thorough study of them. What may be learned I have not now the time to determine, but, from a very cursory glance at them, I learn that Mr. Livingstone's proposition is true. *Small sheep do shear more in proportion to their bulk than large ones,* without regard to age or sex. I learn, also, that yearling ewes shear the largest percentage they ever will shear, and that they shear less and less percentage as they grow older till they are four years old. They gain until they are five, and raising a lamb at that age does not decrease the product of wool, as it has done, but at six they have passed the meridian, and, for the product of wool, commence going ' down hill.'

" It can be seen at a glance that wethers shear their largest percentage when yearlings. At two they have lost 1 per cent., after which they commence gaining, and continue to gain until they are five years old, after which I know nothing of the facts.

" The facts are just as obvious in the classification by weight. The smallest sheep shear the largest percentage, and, as their weight increases, the fleece decreases in proportion till they weigh more than 100 lbs., when it increases the fifth of one per cent., a smaller increase than any decrease in either of the tables. This being the exception to what before seemed to be the rule, leads me to believe that the number in this class is too small, and that I ought to have had 100 sheep at least in this class to arrive at the truth. If it could be ascertained what per cent. of lambs 100 or 1,000 ewes would rear, and the average value of average lambs, it could be very easily calculated which would be the most profitable to keep— a flock of ewes or wethers. But there is no likelihood of this being done, and as ewes are absolutely necessary to increase the flock, perhaps no grazier will be bold enough to keep a flock exclusively of wethers, though I am confident that these tables will prove that the wethers have brought to their owners more money than the ewes.

" If I had the time I might pursue these deductions further, with profit to myself, if not to those who read;

but I think enough has already been disclosed to give any inquiring mind a stimulus to pursue the investigations. Every wool-raiser ought to know which of his sheep he is keeping at a profit and which at a loss."

He ought to discard those sheep that do not shear up to the standard that he has established in his flock, and, keeping only a sufficient number of ewes to supply to his flock the necessary increase, be able to sell each year his five-year-old ewes, and dispose of his wethers when they have reached a suitable age, at the nearest market. A properly proportioned flock consists of one-third breeding ewes and two-thirds wethers.

CHAPTER VII.

DISEASES OF SHEEP.

SCAB.—In sheep, scab, like itch in man, or mange in horses or dogs, depends upon the initiation of a minute acarus, which burrows in the skin, especially if dirty and scurfy, causing much itching, roughness, and baldness.

The parasite readily adheres to hurdles, trees, or other objects against which the affected sheep may happen to rub themselves, and is thus apt to be transferred to the skins of sound sheep.

M. Walz, a German veterinarian who has thrown great light upon the habits of these parasites, says :—

" If one or more female acari are placed on the wool of a sound sheep they quickly travel to the root of it, and bury themselves in the skin, the place at which they penetrated being scarcely visible, or only distinguished by a minute red point. On the tenth or twelfth day a little swelling may be detected with the finger, and the skin changes its colour, and has a greenish-blue tint. The pustule is now rapidly formed, and about the sixteenth day breaks, and the mothers again appear, with their little ones attached to their feet, and covered by a portion of the shell of the egg from which they have just escaped. These little ones immediately set to work and penetrate the neighbouring skin, and bury themselves beneath it and find their proper nourishment, and grow and propagate until the poor animal has myriads of them to prey on him, and it is not wonderful that he should speedily sink.

" Some of the male acari were placed on the sound skin of a sheep, and they, too, burrowed their way and disappeared for a while, and the pustule in due time arose; but the itching and the scab soon disappeared without the employment of any remedies."

By some scab is supposed to be the spontaneous result of bad keep, over heating, and exposure afterwards to great extremes of either wet or cold.

It frequently causes great losses to the sheep-farmers of the mother country. It has been remarked, however, that short-woolled, small sheep, such as the merino, are not so subject to the attacks of the parasite. Referring to this, Mr. Youatt observes :—" The old and unhealthy sheep are first attacked, and long-woolled sheep in preference to the short; a healthy short-woolled sheep will long bid defiance to the contagion, or probably escape it altogether." Regarding post-mortem appearances of an infected sheep, Mr. Youatt further observes :—"The post-mortem appearances are very uncertain and inconclusive. There is generally chronic inflammation of the intestines, with the presence of a great number of worms. The liver is occasionally schirrous, and the spleen enlarged, and there are frequently serous effusions of the belly, and sometimes of the chest. There has been evident sympathy between the digestive and the cutaneous systems."

Remedial applications for this disease are numerous, and consist of arsenical or tobacco dips, and mercurial ointment, for the most part. There are, however, many more elaborate mixtures strongly recommended by various authors, but which are, most of them, at least, unavailable for the Australian squatter, owing to the great and careful manipulation of each sheep which is required for their successful application.

The following is recommended by Mr. Youatt:—"Take common mercurial ointment ; for bad cases rub it down in three times its weight of lard, for ordinary cases, five times its weight of lard. Rub a little of this ointment into the head of the sheep. Part the wool so as to expose the skin in a line from the head to the tail, and then apply a little of the ointment with the finger the whole way. Make a similar furrow and application on each side, four inches from the first, and so on over the whole body. The quantity of ointment (after being compounded with the lard) should not exceed two ounces, and considerably less will generally suffice. A lamb requires but one-third as much as a full-grown sheep."

For all practical purposes in Australia, tobacco will be found the best remedy. It should be boiled in cauldrons, or large iron pots, and transferred to the ordinary colonial contrivance used for sheep-dipping, and the sheep immersed in it in the same manner as they would be if suffering from ticks.

RED WATER, also known as BLOODY URINE, is a disease of cattle, and occasionally of sheep. It depends upon the eating of coarse, indigestible, innutritive food, or continued exposure to inclement weather, and other causes leading to a deteriorated state of the blood. The appetite and rumination are irregular, the bowels speedily become constipated, and the urine reddened by the broken-down globules of the blood. In the more advanced stages the urine is black.

Mr. Spooner is the author of the following description of this disease, and of the proper mode of treating it:—

"The disease understood by this term consists of the effusion of reddish-coloured serum, or water, in the abdomen, outside the bowels, and is the effect of increased action of the membrane called the peritoneum, which forms the outer coat of the bowels, and also lines the abdominal cavity. It is the natural office of this membrane to secrete a watery fluid, in order that the bowels should glide regularly on each other; but when diseased action is set up in this membrane, its secretion becomes excessive, and the serous portion of the blood, mingled with some of the red portion, becomes effused in this cavity, whence it cannot escape.

"The disease is extremely common to lambs, both

during the time they are with their dams and after they
are weaned ; and in them, as well as in sheep, it is very
fatal, destroying the latter in twenty-four hours and the
former in less time.

" The nature of the fluid effused is similar to the serum
or watery portion of the blood ; and as there is no active
pain manifested we are not justified in considering it the
effect of inflammation, but one rather of debility of the
vessels and the existence of too much moisture in the
system.   It usually shows itself in the following symptoms in sheep :—Refusal to feed or ruminate ; a dull,
heavy appearance, often attended with giddiness ; a
staring eye, obstinate costiveness, and sometimes the
head is carried on one side.   In lambs, these symptoms
are less decidedly marked ; but the little animal lags
behind its fellows, is unwilling to move, is very dull, and
dies in a shorter time than the sheep.   Acute pain is
rarely manifested by either sheep or lambs, but they are
generally carried off in a short time."

The treatment of this disease, after the foregoing
symptoms have fully declared themselves, is very seldom
of any avail.   Indeed, among large flocks nothing can be
done but supply plentifully with rock-salt, and this is
recommended more as a preventive than as a cure.

FLUKE, or ROT, consists in the maturation within
the liver and biliary ducts of an entozoon, the " *distoma
hepaticum*," or fluke.   Low, damp, marshy situations
and water meadows furnish a large proportion of
cases.   Sheep grazed even for a few hours upon
land subject to rot, or taking a single draught

from an infected stagnant pool, may contract the disorder, most probably by swallowing young flukes. From fifteen to forty days usually elapse before any serious consequences ensue from the presence of the parasite. At first, indeed, the digestion appears to be stimulated, and the sheep thrive rather better than before; but by-and-bye they rapidly waste, their wool becomes dry and easily detached, their bowels irregular, their skin and mucous membranes yellow, as may be usually conveniently observed by examining the eye and its pearly caruncle, which, in rot, loses the brilliancy of health, and exhibits a dingy-yellow hue. The body, after death, is soft, flaccid, and indifferently nourished; watery effusions are discovered underneath the jaws and in other dependent parts; the small quantities of unabsorbed fat have a dirty-yellow colour; the liver is soft and enlarged, and usually mottled with patches of congestion. In the thick and muddy bile, the fluke, with their myriads of spawn, may be seen floating in variable numbers.

Dr. Harrison observes:—" When in warm, sultry, or rainy weather sheep that are grazing in low and moist lands feed rapidly, and some of them die suddenly, there is fear that they have contracted the rot."

Leeuwenhoeck says that " he has taken 870 flukes out of one liver, exclusive of those that were cut to pieces or destroyed in opening the various ducts. In other cases, and when the sheep have died of the rot, there were not found more than ten or twelve. Then, is the fluke-worm the cause or the effect of rot? To a certain degree, both. They aggravate the disease, and they perpetuate a state of irritability and disorganization which must necessarily undermine the health of any animal. Notwithstanding all this, if the fluke follow the analogy of other entozoa and parasites, it is the effect, and not the cause, of rot."

Mr. Youatt, writing upon this subject, and in reference to the kinds of pastures where this disease is found, and to the season during which it generally assumes its greatest proportions, says:—" It is rarely or never seen on dry or sandy soils and in dry seasons; it is rarely wanting on boggy or poachy ground, except when that ground is dried by the heat of the summer's sun, or completely covered by the winter's rain." There are places where no sheep can be turned to feed with impunity, and others that seldom or never give the rot. The soil of the former either never becomes dry, or takes a long time doing so; that of the latter is so formed as to retain but little moisture. " Some seasons are far more favourable to the development of rot than others, and there is no manner of doubt as to the character of those seasons. After a rainy summer, or a moist autumn, or during a wet winter, the rot destroys like a pestilence. A return and continuance of dry weather materially arrests its murderous progress. Most of the sheep that have been already infected die, but the number of those that are lost soon begins to be materially diminished. It is thus sufficiently plain that the rot depends upon, or is caused by, the existence of moisture. A rainy season and a tenacious soil are fruitful or inevitable sources of it.

The mischief is effected with almost incredible rapidity."
Mr. Youatt further says :—" It is one of the characters
of rot to hasten, and that to a strange degree, the
accumulation of flesh and fat;" and goes on to recommend the sale of these rotted sheep at that point of their
existence when they cease to fatten and begin to make a
retrograde movement. This, however, is a view of the
case which we most strongly condemn. The practice of
selling diseased meat is an infamous one, and is, we
believe, in most countries indictable. Both Mr. Youatt
and Mr. Spooner state that this plan was adopted by the
celebrated Mr. Bakewell; and, if such be the case, all that
he did for the improvement of sheep will not avail to
obliterate this blot on his escutcheon.

The rot is not infectious, and can only be gathered
from herbage or water. If not too far advanced, liberal
supplies of rock-salt, and removal to drier paddocks, will
effect a cure.

From the *Sydney Mail* we quote the following :—

" Nearly twenty years ago (says the *Field*) thousands
of sheep were lost in Great Britain through the ravages
of the parasitic disease known as ' rot,' and in the West of
England as ' lain,' or ' coathe.'

" On wet lands, excepting only marshes near the sea,
rot is constantly present among sheep which are grazed
there; and in wet seasons, when all pasture lands
partake more or less of the marshy character, the
disease prevails to an abnormal extent, as it did in 1860
and 1861, owing to the universal presence of the chief
cause—moisture. Some lands are well known as 'rotti⸴

lands,' and sheep placed on them for a short time exhibit that peculiar tendency to lay on fat which is one of the early stages of the disorder; but after a longer residence on the tainted grounds they suffer from the more fully developed forms of the affection, and then lose flesh as rapidly as they had previously gained it. Rotting meadows are not objectionable as adjuncts to the lairs in which the butcher may keep the sheep he intends to slaughter; but on a sheep farm the smallest piece of ground which possesses the evil reputation is a spot to be avoided, and when discovered, will certainly be fenced off by the flock-master. In 1879, owing to the considerable rainfall, rot was very prevalent in Ireland; according to the *Irish Times*, almost as much so as in the great rotting year, 1860. In various parts of Great Britain the affection also existed to an unusual extent, but not to a degree to excite apprehension.

" Rot, perhaps more than any other well-known disease of the lower animals, has given rise to speculation as to its nature and causes. According to Youatt, the literature of the disease is very ancient. Hippocrates gives a good account of it, and our agricultural writers of the earliest times have described the signs of the affection in terms which leave no room for doubting its identity with that which we are familiar with now. The most remarkable theories have been indulged in by those who have endeavoured to explain the appearance of rot in certain pastures. Particular grasses, which grow in marshy situations, were naturally selected as the cause of the affection by these authorities; others ascribed some

special influence to the gruft or dew; and others have
looked to the innutritious character of the grasses grown
on wet lands as a sufficient explanation of the debility
which is one of the conditions of the advanced form of
the disease.

"Before the true character of rot was ascertained, it
was usual to seek for its causes in the locality where the
sheep happened to be feeding when they manifested
symptoms of illness; whereas the animals had probably
been infected some time before, when their surroundings
were of a totally different character. Notwithstanding
the mystery which formerly attached to the disease, every
one who is learned in the history of sheep is aware that
rot is due to the presence of 'flukes' in the liver, and the
life history of the parasite contains all that is known of
the causation of rot. The fluke worm (*distoma hepaticum*) belongs to the class of trematode worms, which
includes many species and varieties, the *rôle* of each
being to inhabit the liver ducts of different animals, and
thus cause derangement of the digestive functions. To
the Danish naturalist Stienstrup belongs the credit of
having thrown the first ray of light on the darkness
which for a long period had hidden the facts of the
development of the worm.

"In Youatt's time the presence of the worm in the
liver ducts was well known, and the microscopists of
that period had detected the eggs of the parasite in large
numbers in the bile; it was even announced that the
young flukes, just hatched, and not larger than cheese-
mites, had been found in the liver. This last discovery,

however, was purely imaginary, as the researches of Stienstrup, Siebold, Kuchenmeister, and others, have quite established the fact that the fluke is not at once hatched out from the egg in the organism of the sheep, but has to pass through a series of very remarkable changes, from the condition of the ciliated embryo which the egg produces to that of the matured parent.

" In the first place, the creature which emerges from the egg of the fluke is destined to become encysted in the shell of a fresh-water mollusc, then to be parasitic to one of these soft-bodied animals, and in this condition to be swallowed by the sheep along with the herbage or water. These circumstances explain the fact that marshy places are necessary conditions for the existence of rot, as the water-snails, in which the young flukes reside, can only flourish in such situations.

" Sheep which feed in localities where water molluscs abound become the recipients of the inchoate forms of the worm, which ultimately attains the condition of the mature fluke, and in addition they are exposed to the debilitating effects of the insufficient nutriment afforded to them by the unripe and imperfectly grown herbage of the place. How far the development of rot is favoured by the watery food upon which the sheep subsist in many places, it is difficult to calculate ; but there is no doubt that the introduction of the flukes in sufficient numbers into the system will in itself cause the disease. Under ordinary circumstances, however, the sheep is exposed to both causes of disturbance at the same time, and in consequence the progress of the affection is more rapid than it is when the animal, after receiving the fluke larvæ, is removed to a dry, healthy pasture and well fed.

"Rot is not a contagious disease, not even in the imperfect sense in which sheep scab is contagious; both diseases are parasitic. But, while scab can be communicated by the mere transference of the acarus from one sheep to another, flukes cannot be transferred directly to a healthy sheep from one diseased; the eggs of the parasite must be deposited in a favourable situation, the embryo must find a proper habitat, and then pass through several changes of form, each approaching more nearly to the fluke state, until it arrives at the exact point in its development in which it is fit to be introduced into the system of a warm-blooded animal.

"In the course of the generations through which flukes pass in order to attain full development, the sheep or other animal which becomes infected is merely one of the congenial habitats in which they are required to pass a stage of their existence. While in the liver ducts of the warm-blooded animal, the parasites do not multiply, but they attain sexual maturity, and produce millions of eggs, which, being expelled with the bile from the alimentary canal of the infected animal, either shrivel up and die, or give exit to the embryo, according to the nature of the soil and the weather at the time. A rotten sheep, therefore, will be a fruitful source of contamination of land when the conditions are favourable. It will do no harm at all on high and dry soils, or in hot, dry seasons, because in such circumstances the eggs do not

7

advance a stage in development, and probably lose their vitality."

FOOT-ROT among sheep consists of two varieties, the commoner consisting of an inordinate growth of hoof, which, at the toe, or round the margin, becomes turned down, cracked, or torn, and thus affords lodgment for sand or dirt.

Insufficient wearing of the hoof is the obvious cause, and hence the prevalence of foot-rot in soft, rich pastures, and especially among sheep previously accustomed to bare, rough, or upland walks, where the hoof is naturally worn down by the greater amount of walking necessary to procure sustenance. Taken in time, when lameness is first apparent, and before the hoof is cracked and the foot inflamed, a cure rapidly follows a careful paring of the superfluous and diseased portion of the hoof; indeed, further treatment is scarcely necessary, unless any of the vascular parts have been laid bare, when a little tar may be applied as a mild astringent and protection from flies. When, from inattention or neglect, the hoof is separated from the sensitive parts beneath, when ulcers appear on the sole, or proud flesh springs up, active astringents or mild caustics are necessary. The shepherd's old favourite, butyr of antimony diluted with an equal quantity of tincture of myrrh, is a good remedy when cautiously and temperately applied.

A convenient paste, which, in inexperienced hands, is safer than a fluid caustic, may be made with equal weights of flowers of sulphur and finely powdered sulphate of copper, rubbed up to needful consistency with lard or oil. Many have great faith in a mixture of salt of copper with gunpowder and lard.

In reference to this disease Mr. H. S. Randall writes as follows:—"The fore feet are usually first attacked—sometimes both of them simultaneously, but more generally only one of them. The animal at first exhibits but little constitutional disturbance. It eats as is its wont. When the disease has partly run its course in one foot, the other fore foot is likely to be attacked, and presently the hind ones. When a foot becomes considerably disorganized, it is held up by the animal. When another one reaches the same state, the miserable sufferer seeks its food on its knees, and if forced to rise and walk, its strange hobbling gait betrays the intense agony it endures on bringing its ulcerated feet in contact with the ground. There is a bare spot on the under side of the brisket about the size of the palm of a man's hand, but perhaps a little longer, which looks red and inflamed. There is a degree of general fever, and the appetite is dull. The animal rapidly loses condition, but retains considerable strength. Nowhere else do sheep seem to me to exhibit such tenacity of life. After the disappearance of the bottom of the hoof, the maggot quickly closes the scene. Where the rotten hoof is brought in contact with the side in lying down, the filthy ulcerous matter adheres to and saturates the short wool of the shorn sheep, and maggots also are either carried there by the foot or they are speedily generated by the fly. A black crust soon forms, and rises a little higher round the spot. It is the decomposition of the surrounding structures—

wool, skin, and muscle—and innumerable maggots are at work below, burrowing into the living tissues, and literally eating up the miserable creature alive. The black festering mass rapidly extends, and the cavities of the body will soon be penetrated, if the poor sufferer is not sooner relieved from its tortures by death.

" The offensive odour of the ulcerated feet, almost from the beginning of the disease, is so peculiar that it is strictly pathognomonic. I have always believed that I could, by the sense of smell alone, in the most absolute darkness, decide on the presence of hoof-rot with unerring certainty.

" When the malady has been kept under during the first summer of its attack, but not entirely eradicated, it will almost or entirely disappear as cold weather approaches, and not manifest itself again until the warm weather of the succeeding summer. It then assumes a mitigated form: the sheep are not rapidly and simultaneously attacked, there seems to be less inflammatory action in the diseased parts, and less constitutional disturbance, and the course of the disease is less malignant, more tardy, and it more readily yields to treatment. If well kept under the second summer, it is still milder the third. A sheep will occasionally be seen to limp, but its condition will scarcely be affected, and dangerous symptoms will rarely supervene. One or two applications of remedies made during the summer will now suffice to keep the disease under, and a little vigour in its treatment will entirely extinguish it."

Various experiments have proved that the very worst

case of foot-rot can be entirely cured when it makes its first attack upon a flock. But to do this it is necessary to adopt the English methods of applying remedies —namely, daily application of lotion, poultices, &c., and the entire separation of any diseased sheep from its flock companions.

This method is of course quite impracticable among the large flocks of the colonies, and other plans have to be adopted. The separation of the diseased sheep from their healthy companions may be effected, but the separate and single attention to each one of those diseased is quite impossible. In the application of any remedies for the cure of foot-rot the greatest attention must be paid to the hoof-paring part of the operation, and it requires just as much care whether there be ten thousand or only ten units to be performed upon.

We select the following upon the treatment of this disease from Mr. Randall's " Practical Shepherd ":—

" The preparation of the foot is a subject of no dispute, but the labour can be prodigiously economized by attention to a few not very commonly observed particulars. Sheep should be yarded for the operation immediately after a rain, if practicable, as then the hoofs can be readily cut. In a dry time, and after a night that has left no dew on the grass, their hoofs are almost as tough as horn. They must be driven through no mud or soft dung on their way to the yard, which doubles the labour of cleaning their feet. The yard must be small, so that they can be easily caught.

" The principal operator seats himself in a chair—a

couple of good sharp knives (one, at least, a thin and narrow one), a whetstone, the powerful toe-nippers, and such medicines as he may choose to employ, within easy reach. The assistant catches a sheep, and lays it partly on its back and rump, between the legs of the operator, the head coming up to about his middle. If the hoofs are long, and especially if they are dry and tough, they are shortened by the toe-nippers; any filth or dirt between the toes must also be removed, and then, each man seizing his knife, the process of paring the horn is commenced. *And on the effectual performance of this all else depends.*

"If the disease is in the first stage, *i.e.,* if there is merely an erosion and ulceration of the cuticle and flesh in the cleft above the walls of the hoof, no paring is necessary. But if ulceration has established itself between the hoof and the fleshy sole, the ulcerated parts, be they more or less extensive, *must be entirely denuded of their horny covering, cost what it may of time and care.* It is better not to wound the sole so as to cause it to bleed freely, as the running blood will wash off the subsequent applications; but no fear of wounding the sole must prevent a full compliance with the rule above laid down. At most, the blood can soon be staunched, however freely it flows, by a few touches of a caustic— say butyr of antimony.

"If the foot is in the third stage—a mass of rottenness and filled with maggots—the maggots should first be killed by spirits of turpentine, or a solution of corrosive sublimate, or other equally efficient application. It can be most conveniently used from a bottle having a quill through the cork. By continuing to remove the dead maggots with a stick, and to expose and kill the deeper-lodged ones, all can be extirpated. Every particle of loose horn should then be removed, though it take the entire hoof—and it frequently does take the entire hoof, at advanced stages of the disease. The foot should be cleaned, if necessary, with a solution of chloride of lime, in the proportion of one pound of the chloride to a gallon of water."

And now comes the important question, what constitutes the best remedy?

The remedy recommended by Mr. James Hogg, of Scotland, is, turpentine 2 oz., sulphuric acid 2 drachms—to be well mixed before it is used, and freely applied.

Mr. Spooner thinks 1 oz. of olive oil, and double the quantity of sulphuric acid, superior to the above.

Mr. Robert Smith, in his prize essay, gives the following, and tells us he has found it "invaluable, both in staying its progress and curing the disease:—1 oz. corrosive sublimate, 1 oz. blue vitroil, 1 oz. spirits of salts, 1 oz. verdigris, 1 oz. horse turpentine, 1 oz. oil of vitriol, ¾ oz. spirits of turpentine, and 4 ozs. mercurial ointment." This mixture, when not in use, must be kept well secured from atmospheric influences.

Mr. Randall also mentions spirits of turpentine, tar, and verdigris in equal parts; and goes on to say—"Any of these remedies, and fifty more that might be compounded simply by combining caustics, stimulants, &c., in different forms and proportions, will prove sufficient for the extirpation of hoof-rot, *with proper preparatory and subsequent treatment.* On these last, beyond all question, principally depends the comparative success of the applications.

" *First.*—No external remedy can succeed in this malady unless it comes in contact with all the diseased parts of the foot—for if such part, however small, is unreached, the unhealthy and ulcerous action is perpetuated in it, and it gradually spreads over and again involves the surrounding tissues. Therefore every portion of the diseased flesh must be denuded of horn, filth, dead tissue, pus, and every other substance which can prevent the application from actually touching it and producing its characteristic effects on it.

" *Second.*—The application must be kept in contact with the diseased surfaces long enough to exert its proper remedial influence. If removed by any means before this is accomplished, it must necessarily proportionally fail in its effects.

"The preparation of the foot, then, requires no mean skill. The tools must be sharp, and the movements of the operator careful and deliberate. As he shaves down near the quick he must cut thinner and thinner, and with more and more care, or else he will either fail to remove the horn exactly far enough, or he will cut into the fleshy sole and cause a rapid flow of blood. I have already remarked that the blood can be stopped by caustics, but they coagulate it on the surface in a mass that requires removal before the application of remedies, and in the process of its removal the blood is very frequently set flowing again, and this sometimes several times follows the application of the caustic. Cutting down to the crack between the horny and fleshy sole is not enough. The operator must ascertain whether there is any ulceration between the outside horny walls and the fleshy part of the foot, or at the toe, or whether there is even a rudiment of an unreached sinus or cavity in any part of the foot where the ulceration has penetrated or is beginning to penetrate. The practised eye decides these questions rapidly from the characteristic appearances, without the removal of any unnecessary horn; but the beginner must feel his way along cautiously, removing more horn where there is doubt, but so removing it that he will not cause an effusion of blood, or uncover the *healthy* quick, or disarrange the proper bearing of the foot. If the foot is in the third stage, the removal of the maggots, the cleaning of the ulcers, the proper excision of the dead tissues, &c., require much time. The most experienced operator cannot perform these processes in a hurry; the inexperienced one must perform them slowly, or all the time saved will be lost twenty times over, in having to repeat them for an indefinite number of times."

RINDERPEST, or STEPPE MURRAIN, is a contagious, eruptive fever of the bovine species, but sheep, goats, deer, and other allied species occasionally catch it from cattle. It occurs indigenously on the plains of Western Russia, whence it has at various times overspread most parts of the old world. The specific virus from diseased or infected animals is the only source of cattle-plague;

no filth, overcrowding, or other health-depressing cause, has hitherto produced it. An interval of from twelve to fifteen days generally elapses between the time the virus is introduced and the development of the characteristic symptoms. These consist of, according to good authorities, "congestion of the mucous and cutaneous surfaces, with a sort of aphthous eruption, and thickening, softening, and desquamation of the superficial and investing membrane, and run in a tolerably fixed. and definite course, which is not materially altered by any known remedial measures." As a cattle-plague, rinderpest has been known for upwards of a thousand years, and overran the whole of the Roman Empire during the fourth and fifth centuries. In 810 it first visited England, having been received from France, where it was brought by the armies of Charlemagne. Russia appears to have suffered more than any other European country from this disease, although about one and a half centuries ago it destroyed in Holland upwards of 200,000 head of cattle in two years, and about the same time destroyed in Denmark, in four years, 286,000 head, while in some parts of Sweden it spared only two per cent. Throughout Italy it made great havoc, and destroyed 450,000 beasts in Piedmont alone. England was again visited by this plague in 1745, about the same time as it was raging with terrible force in Holland, where it was supposed to have come from. The losses supposed to have been sustained during four years amounted to nearly 300,000 head, although no accurate estimate could be formed. In Nottingham and Lancashire 40,000 died in 1747, and 30,000 perished in Cheshire during six months of 1757.

To show the infectious character of the disease, we may state that it was introduced from Holland to Portsea in a load of hay. Prompt measures being taken, the spread of the disease was stopped, with a loss of £800 worth of cattle. It has several times appeared throughout Europe, and has always been traceable to the vast plains of Russia. This disease may be transferred to healthy animals by inoculation; a small quantity of the nasal or other mucous discharges of a plague-stricken beast, if introduced, develops the disease in a few days.

"So subtle and potent is the plague poison, and so endowed with the power of self-multiplication and growth, that a very minute portion of it finding access to the blood of a healthy animal of the bovine race increases so rapidly that the whole mass of the blood, weighing many pounds, is infected, and every small particle of that blood contains enough poison to give the disease to another animal." Sheep do not take this disease so readily as cattle, and its effect upon them is not so deadly, but they exhibit similar symptoms.

There is no certain cure for rinderpest. The best plan is to destroy the infected animals, and keep all others as carefully separate as possible. In the case of a beast which it is intended to attempt to cure, solid and indigestible food should be withheld, and mashes, gruel, boiled linseed, &c., be supplied, which can be digested without the necessity of rumination. Small and repeated doses of sulphide of soda have in some cases proved useful, and may be administered with moderate doses of ale, whisky and water, sweet spirits of nitre, spirit of ammonia, or strong coffee, carefully regulated. These doses should be drunk by the animal of its own accord in its gruel or mashes. Inhaling chloroform gives temporary relief to the breathing, but does not exert any permanent benefit.

FARCY is a disease resembling glanders in many of its principal features. It is generally produced by similar causes, frequently precedes, and generally accompanies glanders. The imbibing vessels, and glands of the hind limbs, are inflamed and tender; they become much swollen, hard, and knotted. The impaired animal fluid softens, and is then poured out. This is succeeded by ulcers, or farcy buds; these are very similar to the ulcers of glanders, but, unlike them, they are curable with the exercise of great care.

To prevent their spreading, a searing-iron should be used and the ulcers well scarified. The animal afflicted must be well fed and comfortably lodged.

GLANDERS is a very fatal and quite incurable disease. It is much more common among horses than any other animal, but it frequently attacks sheep, and sometimes human beings. It makes its first appearance in the shape of small holes, or ulcers, in the nostrils of the animal, the edges of which are very inflamed or rugged, and the substance they exude is of a greenish, sticky, offensive nature, with a great inclination to spread. The blood of the diseased animal becomes thin and weak, and is inadequate to support the system, and the body consequently becomes thin and wasted. The mucous membrane being insufficiently nourished, causes a nasty cough and open bowels. The lymphatic vessels and glands also become inflamed and swollen, and can be distinctly seen and felt beneath the throat and jaws, and sometimes in the limbs, where they form themselves into ulcers, presenting the appearances of farcy.

This disease, so malignant in its character, is produced by many causes, among which may be mentioned (among horses) hard work, bad feed, close confinement, or any other system of treatment calculated to produce a deteriorated state or condition of the animal's blood. Horses have been known to catch this disease from merely breathing, for a short period of time, a close or impure atmosphere. A very small portion of the discharge from a glandered beast is sufficient, if placed in contact with a cut or abrasion on the skin of a human being, to communicate the disease, which generally terminates fatally. In this way many stable attendants have died. The English Government, in 1853, made it compulsory to destroy any affected animals as soon as glanders made its appearance. Cattle and dogs are not subject to this disease.

Although the diseases farcy and glanders are more subject to the equine species than sheep, the latter sometimes become infected with it; and but little can be done to remedy the disease. We have not seen, or heard of, any instances of this malady among sheep in the colonies; and it is to be hoped that, in future, as great an immunity from it may exist as has done in the past.

TAPEWORMS present, in their mature state, a globular appearance, with a very thick shell; their diameter is about $\frac{1}{864}$ of an inch. Their embryo, upon being transferred to the intestinal organs of the animal, immediately escape from their shell, and bore their way through the living tissues, and lodge themselves in the fattest part of the flesh, there to await a further transformation.

The animal thus afflicted becomes measled, and the embryo becomes transformed into the *Cysticercus cellulosæ.*

It is not definitely known what length of time a tapeworm can exist in an intestinal canal; but it is more than probable that there is some definite period at which the parasite separates from the mucous membrane of the sheep, and when this separation takes place the worm is expelled in the same manner as if the parasite had been killed medicinally.

Leuckart has " proved by experience that the measles, or cysticerci, which produced this worm are to be found in the muscles and internal organs of cattle. He administered proglottides of tapeworm (*mediocamellata*) to three calves, a sheep, and a pig. In the two last-named animals they produced no effect, as was seen by their post-mortem examination; whilst in the calves they produced a kind of leprosy, which has since been characterized as acute cestoid tuberculosis, and which proved fatal if too large a dose of eggs was administered.

On examining one of these animals after its restoration to health (forty-eight days after the eggs were swallowed), he found numerous cysticerci—vesicles larger and more opalescent than those of the pig—lodging within the muscles; and as the heads of the contained cysticerci exhibited the distinct peculiarities presented by the head of the adult worm, we are supplied with the most unequivocal evidence that man becomes infected by the second form of tapeworm by eating imperfectly-cooked meat."

Although tapeworms seldom afflict cattle and horses, they are common in sheep, and, by irritation of the bowels, cause them to present a most unthrifty appearance. The only applicable cure for this disease amongst our Australian flocks is, as is the case with fluke, an abundant supply of rock-salt; but where small flocks are kept, and the sheep housed and hand-fed, the following will be found an effectual cure :—Forty drops of oil of turpentine, one ounce of common salt, one drachm of powdered green vitriol, given in a little gruel ; or, when the bowels require opening, linseed oil may be substituted for gruel.

SHEEP-LOUSE, or SHEEP-TICK, is an insect belonging to the same family as the forest fly. It lives among the wool of sheep, and differs from its family companions in having no wings. It is found particularly in the wool of lambs, and lives by sucking the blood of the animal, and is most abundant in the early part of summer. Where it fixes its head on the skin a large round tumour is formed. Its body is very compressed and smooth, of a rusty colour; the head and thorax small, the abdomen large.

The female does not lay eggs, but, like the other *Hippoboxidœ*, hatches the egg and nourishes the larva within her own body till it passes into the pupa state, when it is deposited, oval-shaped and shining, fastened to the wool of the sheep. Sheep graziers use many washes or *dips* for the destruction of these creatures, many of which are arsenical. A patent was obtained a few years ago for a sheep dip of which carbolic acid is a principal ingredient.

BLAIN.—Mr. Youatt gives the following description of this disease, its symptoms and treatment :—

"Sheep are liable, though not so much as cattle, to that inflammation of the tongue, or rather of the cellular tissue on the side of and under the tongue, to which the above singular name is given. A few sheep in the flock are occasionally attacked by it, or it appears under the form of an epidemic.

"A discharge of saliva runs from the mouth, at first colourless and devoid of smell, but soon becoming bloody, purulent, and stinking. The head and neck begin to swell, and the animal breathes with difficulty, and is sometimes suffocated. A succession of vesicles have arisen along the side of the tongue—they have rapidly grown—they have broken—they have become gangrenous—they have formed deep ulcers, or deeper abscesses, that occasionally break outwardly. When this is the case, it is probably the 'Greathead' of Mr. Hogg. The cause is some unknown atmospheric influence, but the sheep have been predisposed to be affected by it, either by previous unhealthy weather, by feeding on unwholesome herbage, or by unnecessary exposure to cold and wet.

"Whatever may be the case with regard to cattle, there is no doubt that the blain is often infectious among sheep. The diseased sheep should be immediately separated from the rest, and placed in a somewhat distant pasture.

" The malady must first be attacked locally. If there are any vesicles in the mouth, they must be freely lanced. If any tumours appear upon the neck or face, and that evidently contain a fluid, they must be opened. The ulcers must be bathed with warm water at first, and until the matter is almost evacuated; then lotions of cold water, in each pint of which one drachm of chloride of lime has been dissolved, must be diligently used. Aperients must be administered very cautiously, and not at all unless there is considerable constipation. The strength of the animal must be supported by any farinaceous food that it can be induced to take.

" Bleeding will be very proper in this disease, before the vesicles have broken or the external tumours begin to soften, and there is an evident and considerable degree of fever ; but after the purulent, fetid matter has begun to appear, it will only hasten the death of the animal."

Hoove, a disease common enough in small holdings, in any part of the world, is thus spoken of by Mr. H. S. Randall :—" When sheep are suddenly turned from poor pastures on fresh clover, turnips, or other unusually succulent food, and allowed to fill themselves to excess, its fermentation in the first stomach or paunch causes an elimination of gas, which sometimes distends that organ almost to bursting. It presses against the diaphragm (midriff) so that the lungs cannot be filled with air, and thereby directly produces suffocation, or the blood no longer circulates through the paunch, and is determined to the head, producing stupor and death.

"It is a most egregious folly in all cases to make any such change in food. If dried-off ewes, for example, are to be put on rank clover, they should at first be admitted to it for only two or three hours a day, and driven in at midday, when their hunger is to a considerable extent satisfied. This continued for two or three days entirely prevents the danger of hoove.

"When sheep are discovered to be 'hooven,' they should be driven gently about for an hour. If swollen to a very dangerous extent, and the distress and oppression are evidently increasing, they must be relieved by mechanical means. Those possessed of such instruments either pass a flexible tube with a rounded perforated end down the throat into the stomach, through which the pent-up gas escapes, or they plunge a trocar (a sharp stylet or puncturing instrument, covered with a canula or sheath) into the paunch through the left flank. The trocar is withdrawn, leaving its sheath in the wound, which keeps the openings in the side and paunch opposite to each other, thus allowing a freer exit to the gas, and preventing the other matters forced along with it from being left in the cavity of the abdomen or belly. Any solid or semi-solid matter deposited there leads to inflammation, and ultimately, if in any considerable quantity, to death. If a pocket-knife is used instead of a trocar, the above dangers are increased, but it is often the only available instrument at hand, and generally proves a safe one. The place for inserting it is in the left flank, half-way between the haunch and ribs, and well up towards the back-bone."

Purgatives are, however, considered the safest remedial measures. Mr. Spooner prescribes—"Sulphate of magnesia, 2 ozs.; ginger, 1 drachm; gentian, 2 drachms; chloride of lime, ½ drachm; to be dissolved in a pint of warm water."

Mr. Randall continues—"If gas continues to be developed, Mr. Youatt recommends the introduction into the stomach of chloride of lime—a drachm dissolved in a gill of water—either by means of a horn, or through the canula of the trocar. This would also be an admirable remedy to administer (down the throat) in earlier stages of the disease, when the case was not urgent or the opening of the paunch yet called for. Once in the paunch, it would produce chemical results which would at once relieve the parts of their unnatural distension."

ENTERITIS, or inflammation of the bowels, and especially of their muscular and serous coats, is a disease common not only to the various breeds of lower animals, but to man. Among sheep it is generally produced by undue privations, or improper pasturage, such as that containing very coarse, wet grass, or acrid or poisonous plants. It also generally appears among flocks depastured in cold, wet localities. The symptoms are fever and thirst, a quick but rather weak pulse, restless twitching of the hind legs, tenderness of the belly, and torpidity of the bowels.

Mr. Youatt recommends "bleeding according to the age and condition of the animal, and the urgency of the symptoms; purgatives perseveringly administered until the bowels are opened, and the purging being afterwards kept up, the epsom salts being employed to produce the first effect, and sulphur the second. The food to consist of mashes or gruel. No tonic to be allowed until the febrile stage is passed, or until violent diarrhœa has succeeded to the constipation."

This will be found quite effectual for horses, cattle, and sheep. Indeed, in the case of the latter, it can only be used under peculiar circumstances, such as a very limited number of sheep requiring treatment, or some particular reason for their preservation which would render a flock-master more solicitous than these gentlemen generally are in Australia, where, in general, a few hundred sheep, more or less, are considered immaterial.

BRAXY.—The vague way in which the term braxy is used renders it difficult to define the disease, for, in different parts of the world, totally different diseases are included under this head. Of the two most generally recognized as braxy, the one is an intestinal affection, attended with obstinate diarrhœa; the other is a blood disease, and the result of plethora or fulness of blood. The first of these we will treat upon under the heading of diarrhœa, and the second, which is considered by good authorities to be the true braxy, we will describe here.

A very lean flock of sheep, suddenly placed upon rich food, is apt to be decimated by braxy. Hilly land is favourable to the production of braxy, from the firm nature and nutrient qualities of grass growing upon it. We find the disease in such situations in the winter season. In Great Britain, about November, many of the well-fed hogs, placed upon turnips, die suddenly from braxy; and, again, when farmers resort to the forcing system towards spring, the mortality is great, particularly when, in addition to much artificial food, sheep are allowed rich pastures.

The mortality is greatest during the full moon, from the sheep grazing during the light night as well as by day. The old country shepherds at these times very frequently find one or two dead in the morning. Some assert that, in the winter, exposure produces braxy; and it is very possible that it may be produced by any sudden checks to the exhalations, which tend so much to maintain the balance of the functions and purify the blood.

The sheep attacked with braxy exhibits the following symptoms:—In full health at the time, he suddenly ceases to eat, has a staring look, is peculiarly excitable, and separates himself from the flock. The head is lifted high, the breathing becomes laboured, the countenance appears anxious, and the animal loses the power of his limbs. He totters, falls over, is seized with convulsions, and dies within five or six hours, and often within one hour, from the appearance of the first symptoms of the disease.

If the sheep's throat be cut before it dies, the absence of any peculiar appearance within the body after death is very remarkable. The flesh appears of a dark-red colour, and the veins are charged with dark blood, but, on the whole, the body presents a good appearance. On the other hand, if the throat is not cut before death, the carcass rapidly becomes distended, decomposed, and stinking.

The treatment for braxy is altogether preventive, and it consists in the regulation of the animal's diet, to prevent sudden transitions from low, or poor, to rich food.

Occasional doses of nitre, or salts, may be administered if practicable.

DIARRHŒA.—This disease, as already remarked, is one form of braxy, and consists of an increase in the discharge from the bowels. These discharges are usually unduly liquid. It is closely allied to dysentery, and often in the after stages takes that form of disease. Mr. Randall, speaking of it, says :—

"Common diarrhœa, purging or scours, manifests itself simply by the copiousness and fluidity of the evacuations of dung. It is brought on by a sudden change from dry feed to green, or by the introduction of improper substances into the stomach. In diarrhœa there is no apparent fever, the appetite remains good, the stools are thin and watery, but unaccompanied with mucus (slime) and blood. The odour of the dung is far less offensive than in dysentery ; the general condition of the animal is but little changed."

In the cases of grown sheep it is, perhaps, unnecessary to attempt any remedy, as but few cases occur with fatal terminations. Lambs, however, if attacked at certain periods of the year, require some attention—that is, among such flocks as it is found practicable to treat sheep medicinally—frequently. Mr. Randall recommends, for lambs, " half a drachm of rhubarb, or an ounce of linseed-oil, or half an ounce of epsom salts. This should be always followed by an astringent, and in nine cases out of ten the latter will serve in the first instance."

Mr. Spooner says :—" If the cases are not severe, and are entirely confined to diarrhœa, astringents alone may be given; but, if any mucus is perceived, it will be proper to administer a laxative in the first instance."

DYSENTERY.—This disease is attended by discharges from the bowels, and differing from diarrhœa chiefly from being attended by marked fever and pain, as also by the presence of blood and inflammatory products in the discharges. On dysentery, we quote from Mr. Randall's " Practical Shepherd " as follows :—" It differs from diarrhœa in several readily-observed particulars. There is evident fever ; the appetite is capricious, ordinarily very feeble ; the stools are as thin, or even thinner, than in diarrhœa, but much more slimy and sticky. As the erosion of the intestines advances, the dung is tinged with blood, its odour is intolerably offensive, and the animal rapidly wastes away. The course of this disease extends from a few days to several weeks." According to Mr. Randall's recollection, about one-third of the cases that came under his notice proved fatal, but they were usually old and feeble sheep that succumbed.

For the treatment of this disease Mr. Youatt recommends bleeding.

CESTOID WORMS.—The number of different kinds of cestoid worms is great. Their natural history is important, not only in regard to the health of sheep and other valuable domesticated animals, but also in reference to that of human beings. Cestoid worms in their most perfect state, when alone they possess the form from which their name is derived, are in reality compound animals. The head of the cestoid worm is furnished with organs—different in different kinds—by which it affixes itself to the inner surface of the intestine of a vertebrate animal. When first it gets into this situation the body is very short, and has no joints; but they soon begin to appear, and, gradually increasing in size, become, in most of the kinds, very distinct, and at last separate from the system in which they were produced and are carried away out of the intestines of the animal which contained them. This does not take place, however, till they have not only become mature in the development of the sexual organs—the principal organs to be observed in them—but until they are full of what are called eggs, which, indeed, are rather young ones ready for a separate existence, and each developed in a sort of protective shell. Each joint of a cestoid worm is androgynous. Whilst the most matured joints are thrown off from the posterior end, new joints are continually formed, as at first, in the part nearest to the head.

The number of joints thus formed from a single individual is very great, as will appear when it is considered that tapeworms have been found twenty feet long or upwards, and that these have probably been throwing off joints in large numbers before opportunity has been obtained of measuring them.

In some cestoid worms the division into segments is imperfect. When segmentation is perfect, the segments, in separating from the parent system, possess life, and some power of independent motion, creeping away on moist ground, plants, &c. Their period of separate existence, however, is brief; they burst or decay, and the numerous minute embryos which they contain are ready to commence their career if in any way transferred into the stomach of an animal of proper kind, which is generally different from that whose intestine their parent inhabited. This may happen by their being swallowed along with grass, water, &c. Some find their appropriate places in the stomachs of vertebrate and others in those of invertebrate animals.

The shell being broken or digested, the young cestoid worm is set free. It is generally unlike the proglottis by which it was generated. It presents the appearance of a vesicle provided with a few microscopic hooks. It possesses, however, a power of active migration, by means of these hooks, and is able to perforate the stomach of the animal which contains it. To this its instinct seems immediately to prompt it, and it is so minute that it passes through the stomach without causing any serious inconvenience to the animal. It now probably gets into the blood, and is lodged in some of the capillaries, from which it makes its way again, by perforation, until it finds a suitable place in some of the tissues, or of the serous cavities in the flesh, or in such organs as the liver or the brain, and here, relinquishing all active migration, it rapidly increases in size, at the same time developing a head, which is, in fact, that of a cestoid worm, and generally either encysts itself or is encysted (enclosed in a cyst), according to circumstances, or according to its species.

Great numbers of such parasites are sometimes found in a single animal, causing disease, and even death. In reference to intestinal worms, Mr. Randall remarks that he has heard of but one single alleged fatal case in the United States of America; and Mr. Spooner says:—— " Sheep are subject, in rare instances, in England, to a disease arising from the presence of worms in the intestines. Mr. Copeland, of Suffolk, found fifty lambs labouring under violent diarrhœa. On examining some which died, he found large patches of inflammation on the villous membrane of the fourth stomach. The small intestines contained thousands of the folded tapeworm, and about twenty-five of the large round worms, with a large quantity—several ounces—of sand."

HYDATID.—True hydatids were formerly regarded as cystic entozoa; but all such animal forms are now discovered to be larval stages of tapeworm. These hydatids may occur in almost any part of the body, and they have been discovered in man, the ape, the ox, the sheep, the horse, the camel, the pig, the kangaroo, and some other vegetable-feeders; but they apparently do not occur in carnivorous animals, nor in the rodents. They are generally enclosed in an external sac, which is attached to the tissue of the organ in which it is situated, and which is frequently common to many hydatids, each of which has a distinct envelope. The fluid in the interior of the hydatid itself is almost colourless and limpid; but the fluid in the common cyst in which the hydatids float is sometimes of a yellow colour.

Hydatids are found in the brains of various ruminants, and give rise to the disease in sheep known as the staggers. When the hydatids occur in the fourth ventricle, the animal, instead of turning round and round in one direction, springs in the air; and this variety of the infection is hence distinguished by German veterinarians as *Das Springen*.

Hydatids sometimes occasion so little inconvenience that their presence is not discovered until after death. On other occasions they grow so rapidly, and cause so much irritation, that suppuration is excited in or around the common sac, which may either burst externally, or into a mucous canal or a serous cavity. In the first and second cases the hydatids will be discharged, and recovery may take place; in the third case, fatal inflammation will ensue.

Mr. Spooner says the symptoms are—" a dull, moping appearance; the sheep separating from the flock; a wandering and blue appearance of the eye, and sometimes partial or total blindness; the sheep appears unsteady in its walk; will sometimes stop suddenly, and fall down; at others, gallop across the paddock; and, after the disease has existed for some time, will almost constantly move round in a circle (or spring in the air, as before mentioned). There seems, indeed, to be an aberration of the intellect of the animal. These symptoms, though rarely all present in the same subject, are yet sufficiently marked to prevent the disease being mistaken for any other."

According to Mr. Youatt, weakly lambs are very liable to attack. He remarks :—" It usually appears during the first year of the animal's life, and when he is about, or under, six months old; and generally succeeds a severe winter and a cold, wet spring." He further remarks :—" This is a singular disease, but it is a sadly prevalent and fatal one in wet and moorish districts. It is much more fatal in France than in Great Britain. It is supposed that nearly a million of sheep are destroyed in France every year by this pest of the ovine race. The means of cure are exceedingly limited. They are confined to the removal or destruction of the vesicle. Medicine is altogether out of the question here."

If the cyst is near the surface, as sometimes happens, it may be punctured. Mr. H. S. Randall, after alluding to the various methods of puncturing for the cure of this disease, remarks, speaking of America (and as his expressions are equally applicable to Australia, we quote them): —" But when we take into consideration the hazard and cruelty attending the operation, at best, and the conceded liability of a return of the malady—the growth of new hydatids—it becomes apparent that, in this country, it would not be worth while, unless in the case of uncommonly valuable sheep, to resort to any other remedy than depriving the miserable animal of life."

GRUB IN THE HEAD.—We have never seen an instance of this malady. We quote regarding it from Mr. Randall :—

"In the months of July and August sheep are often seen gathered in dense clumps, with their heads turned inwards and their noses held down to the ground. If driven away, they run without raising their heads, or rapidly thrust them down again, as if they had some urgent motive for retaining them in that position. Occasionally they stamp or strike violently with their fore feet near their noses, as if an enemy, invisible to the spectator, were assailing them at that point. It is the *œstrus ovis,* or gad-fly of the sheep, attempting to deposit its eggs within their nostrils." In relation to this, we may remark that Mr. Randall refers to America, and that the months of January and February would, in the Australian colonies, be synonymous with those he mentions. Mr. Youatt says :—" The head and corslet of this insect, taken together, are as long as their body ; and that is composed of five rings, tiger-coloured on the back, with some small points and larger patches of deep brown colour. The belly is of nearly the same colour, but has only one large circular spot in the centre of each ring. The length of the wings is nearly equal to that of the body, which they almost entirely cover. They are prettily striped and marked." Mr. Randall continues :—" If the fly succeeds in depositing its eggs within the nostrils of a sheep, they are immediately hatched by the warmth and moisture, and the larvæ, or young grubs, crawl up the nose, finding their way to the sinuses, where, by means of the tentacula, or hooks, which grow from the sides of their mouths, they attach themselves to the membrane lining those cavities, and there remain feeding on its mucus until the following year. As the minute worm ascends the nose, the sheep appears to be distracted with apprehension. It dashes wildly about the paddock, stamping, snorting, and tossing its head.

" Some farmers protect their sheep measurably from the attacks of the *œstrus ovis* by ploughing a furrow or two in different portions of their pastures. The sheep thrust their noses into this on the approach of the fly. Others smear their noses with tar, or cause them to smear them themselves by sprinkling thin salt over tar."

WATER ON THE BRAIN.—This disease we have had no personal experience of, nor have we heard of any well-authenticated case in the colonies. It has, however, to our knowledge, been sometimes mistaken for sturdy, a malady elsewhere treated upon under the heading of Hydatid.

Mr. Youatt has described water on the brain as " an effusion of serous fluid or water, without being confined in any sac or bladder, within the cavity occupied by the brain, or between its investing membranes. It is peculiar to young lambs, and sometimes occupies the head before birth, giving it unusual size, and rendering parturition difficult. The skull is a little enlarged ; the bones of it are generally thin, but sometimes they are thickened. The appetite occasionally fails, but oftener increases; the bowels are usually constipated, though sometimes they are relaxed; the lamb appears more or less stupified, is disinclined to move, staggers slightly, pines away almost to a skeleton, and dies before it is two months old."—*Randall's "Practical Shepherd."*

This disease is considered generally incurable.

APOPLEXY.—This disease is, we believe, quite unknown in the Australian colonies. No instance of it has ever been brought to our notice. It is, however, common among the improved mutton breeds of Great Britain, and is caused by the system adopted of forcing the sheep forward to the greatest attainable growth, by the use of rich, artificial, stimulating food.

The treatment, in such cases as treatment may be of any avail, is bleeding from the jugular vein until the animal shows signs of weakness.

INFLAMMATION OF THE BRAIN.—On this subject we again quote from Mr. Randall:—" This is a secondary effect of the causes which produce apoplexy, by which the substance of the brain, its membranes, or both, become the subject of inflammation. The symptoms are much more violent than the preceding. After a degree of dulness and inactivity, accompanied by redness and protrusion of the eyes, the animal becomes delirious, rushes about the paddock, attacks men and trees; 'and,' says Mr. Spooner, ' in lambs, their motions are quite ridiculous, and have, in consequence, among the ignorant, given origin to the idea of their being bewitched.' The disease is treated in the same way with apoplexy."

TETANUS, or LOCKED JAW, is a disease common to most of the domesticated animals, but more particularly to horses and sheep. It is usually produced by exposure to cold and wet, by intestinal worms, obstinate constipation, or injuries. The symptoms usually come on gradually; involve, tolerably equally, most of the muscular structures, which become hard and rigid; the nose is protruded, the limbs move stiffly, the tail is upraised, the bowels are constipated. Purgative medicines are recommended. In adult animals most cases are fatal; but amongst young animals, especially when the attack results from exposure to cold, many recoveries occur.

EPILEPSY.—Mr. Youatt remarks that " tetanus and epilepsy may be regarded as kindred diseases in all animals, but in none do they assimilate to each other as in the sheep." The animals, when attacked, writhe with involuntary spasms, fall to the ground, and are for the time without sight or hearing. Sometimes the muscles of the throat are so involved that fatal suffocation ensues. The attack is generally preceded by dulness, and lasts from ten to thirty minutes. In some parts of Europe this malady has proved so fatal that sheep husbandry has been given up. A French writer (Tessier) attributes it to the nature of the pasture. It is also very common in England, where it is believed to arise from an over-high condition. Opening medicines are strongly recommended.

PALSY. — Of this affection we have never had any experience. Mr. Randall speaks of it as " a suspension of the nervous influence on the muscles—the opposite of tetanus and epilepsy, which excite them to unnatural action. The sheep sometimes becomes powerless in every limb, and unable to move ; sometimes the palsy extends only to the loins or hindquarters. It is produced by cold and improper exposure. Young lambs, when weaned in very cold weather, are most subject to its attacks—though grown sheep are not exempt from them ; 'and particularly,' says Mr. Spooner, ' the ewe that has aborted, or produced her lamb with difficulty, and after a tedious labour, in cold weather.' "

PINING. — The disease known by this name is thus spoken of by Mr. Hogg :—" The distemper is a strange one ; it may affect a whole flock at once. The first symptoms are lassitude of motion, and a heaviness about the pupil of the eye, indicating febrile action." The malady is supposed to arise from " an enervated and costive habit, producible by want of proper exercise and eating astringent food."

" The lands which are now most subject to this disease were once, in the same manner, liable to rot. As the draining of the sheep pasture proceeded, the rot gradually became extinct, and was ultimately superseded by the pining."

If it be not possible to remove sheep, suffering from attacks of this distemper, to other pasture of different character, they will probably die ; and Mr. Hogg says, " all those affected will fall in the course of a month."

INFLAMMATORY FEVER.—From the "Practical Shepherd" we learn, on this subject, that " Mr. Price, an English

9

writer, gives the following account of it :—The number of animals that die of this disorder in Romney Marsh is truly astonishing. My opinion is that the soil of Romney Marsh, being very rich (consequently the clover and grasses equally so), that sheep feeding on these rich pastures must be more subject to inflammation than those fed on poorer soils, particularly in the spring, when the young shoots of the grasses and natural clover are full of juices.

"It is said that bad-mouthed sheep never die of this disease, because they cannot feed on short, nutritious grass, but on long, coarse herbage, which does not enrich the blood."

MALIGNANT INFLAMMATORY FEVER.—Mr. H. S. Randall says :—" This malady appears occasionally in England, but is common as a very destructive epizootic in France, where it is termed *la maladie de soligne*. It prevails in low, marshy districts, where the sheep are wintered very poorly and turned out in spring to gorge themselves on the watery, rapidly-growing vegetation. It appears towards the close of spring, and rages until August." Its early symptoms are, according to Mr. Youatt—" Suspension of rumination, loss of appetite, dulness, weeping from the eye, coldness of the ears, alternate shiverings and flashes of heat. Soon after the mouth and the breath become hot, the eyes are red, the pulse is accelerated and weak and irregular, and there is a mucous discharge from the nostrils, to which succeeds bloody mucus, and then a mixture of purulent matter and blood. By degrees the urine becomes bloody, and the excrements are covered with grumous blood. The head and legs are swelled, the debility is extreme, and the animal dies in the course of eight or ten days.

"The greater part of the animals attacked by this disease perish. The sheep in the finest condition die soonest, and with greatest certainty."

TYPHUS FEVER.—Of this malady we have never seen an instance, nor have we met with anyone with more extended experience regarding it.

Mr. Youatt is of opinion that many sheep die yearly in Great Britain from its ravages, and believes that many ailments in sheep recognized as braxy come properly under this heading.

CATARRH, a disease common to man as well as the lower animals. It is commonly looked upon as only a cold. It is, however, in reality, a more definite and clearly displayed ailment than that. It consists of inflammation of the mucous membranes of the nose, and consequent discharge of mucus.

In its mild forms it is termed snuffles, and Mr. Randall tells us that "high-bred English mutton sheep, in this country (United States), are apt to manifest more or less of it after every change of weather."

Catarrh is dependent upon climatic influences, and the Australian colonies are not much likely to be greatly troubled with it, requiring, as it does, sudden falls of temperature, and other epidemic or atmospheric causes. It is common, however, in moist, changeable, temperate latitudes. But few cases prove fatal, and if the sheep have sufficient and proper pasturage but few will be attacked, and none are likely to die from the ravages of catarrh.

MALIGNANT EPIZOOTIC CATARRH.—The term epizootic is applied to diseases of animals which manifest a common character and prevail at the same time over considerable tracts of country. Like epidemics, they appear to depend upon some peculiar and not well ascertained atmospheric causes.

Epizootic catarrh, in the years 1846 and '47, made great ravages among the sheep of the graziers in the state of New York, some losing as many as seven-eighths of their entire flocks. The symptoms, as described by Mr. Randall, are :—" The countenance was exceedingly dull and drooping, the eye kept more than half closed, the caruncle, lids, &c., almost bloodless. A gummy, yellow secretion below the eye; thick, glutinous mucus adhering in and about the nostrils; appetite feeble, pulse languid, and the muscular energy greatly prostrated. Nothing unusual was noticed about their stools or urine."

PNEUMONIA, OR INFLAMMATION OF THE LUNGS.—From the " Practical Shepherd " we select as follows :—" The adhesions occasionally witnessed between the lungs and pleura of slaughtered sheep betray the former existence of the disease. The sheep labouring under pneumonia is dull, ceases to ruminate, neglects its food, drinks frequently and largely, and its breathing is rapid and laborious. The eye is clouded, the nose discharges a tenacious fetid matter, the teeth are ground frequently, so that the sound is audible to some distance. The pulse is at first hard and rapid, sometimes intermittent, but before death it becomes weak. During the height of the fever the flanks heave violently. There is a hard, painful cough during the first stages of the disease. This grows weaker, and seems to be accompanied with more pain, as death approaches."

BRONCHITIS.—Mr. Randall says :—"It would be diffi-cult to suppose that where sheep are subject to pneumonia they would not also be subject to bronchitis, which is an inflammation of the mucous membrane that lines the bronchial tubes, or air passages of the lungs."

PLEURISY, OR PLEURITIS.—Mr. Spooner thus speaks of this malady :—"This disease consists of inflammation of the pleura, or membrane lining the chest. It is produced by the same causes as inflammation of the lungs, with which it may be accompanied, and particularly by any sudden changes that may chill the whole system. It often occurs from this cause after sheep-washing, when it is very common to find a few sheep failing, and in proportion to the want of care exercised. It is not unusual, in examining the bodies of sheep, to find the lungs in part adhering to the sides of the chest, and the animal thus affected generally loses flesh. This adhe-sion is the result of pleurisy, and another and more dangerous effect is water on the chest.

" The symptoms of this disease are in many respects like those of inflammation of the lungs, but are attended occasionally by severe pain, and by a variation of the symptoms generally, such as a harder and more defined pulse, and more warmth of the body."

CONSUMPTION.—Of the lower animals, sheep and oxen

are more subject to this disease than most of the others. In the case of horses, for instance, the health-depressing influences which produce it in man and other animals, induce in them glanders and farcy. It is also rare among dogs, but extends, most commonly of all, to pigs.

It is one of the chief causes of death among the apes and other denizens of our zoological gardens. It is notoriously hereditary, is frequently developed by breeding from parents closely relative to one another, and in this respect, as well as in the causes that produce it, this disease among the lower animals is similar to that which attacks man. We have never seen an instance of this disease among Australian flocks, and quote from Mr. Youatt regarding it:—"There is still another and more frequent and equally fatal disease of the lungs (with acute inflammation), but it assumes an insidious character, and is not recognized until irreparable mischief is effected—viz., sub-acute, or chronic inflammation of the lungs, and leading on to disorganization of a peculiar character—tubercles in the lungs—and terminating in phthisis (consumption). The sheep is observed to cough. He feeds well, and is in tolerable condition. He is driven to market and slaughtered, and the meat looks and sells well; but in what state are the lungs? Let him who is in the habit of observing the plucks of the sheep as they hang by the butcher's door answer the question. He sees plenty of sound lungs from oxen; he sees the lungs of the calf in a beautifully healthy state; but he does not see one lung in three of the sheep that is unscathed by disease—whose mottled surface does not betray inflammation of the investing membranes, and in the substances of which there are not numerous minute concretions—tubercles.

" This disease is especially prevalent in low and moist pastures, and it is of the most frequent occurrence in spring and in autumn, and when the weather at these seasons is unusually cold and changeable. It is almost useless to enter into the consideration of treatment. It would consist in a change to dry and wholesome and somewhat abundant pasture, and the placing of salt within the reach of the animal."

ABORTION.—This may arise from various causes. Any violence, such as a kick from a horse or other animal, or a blow from the horns of a ram, may produce it. Even careless or rough usage at the hands of an operator for foot-rot, or, more likely still, of a cruel shearer, will cause it. It is a great injury to a ewe to abort. She, after doing so, generally becomes weak and poor, her wool ceases to grow, and may have a tendency to drop off, and she fails to recover her former condition for many months. It is a mishap not likely to occur in Australia, except from any of these causes we have mentioned, or from others of a similarly violent or accidental nature.

Mr. Youatt ascribes it to the supply of salt in England, but Mr. Randall differs from him ; and from what we have seen of continued supplies of salt to sheep, and its effects, we do not consider it by any means a probable cause.

Mr. Spooner, again, believes it to arise from too great a quantity of succulent food; but here, also, Mr. Randall is of a different opinion, and has not known of any cases which have arisen from that cause.

The lamb, when born, is almost always dead.

GARGET, OR INFLAMED UDDER.—From Mr. Randall's "Practical Shepherd" we gather that "high-fed English ewes are subject to an acute and dangerous form of this malady. Hard kernels or tumours form in the udder. The udder itself is much swollen and inflamed, with great heat and tenderness. If matter forms in any part of the udder, an incision should be made and the pus squeezed out."

This sometimes arises from the ewe having lost her lamb; and, under these circumstances, good authorities recommend fomentation—a form of treatment not likely to be much adopted by Australian flock-masters.

PARTURIENT OR PUERPERAL FEVER.—According to Mr. Randall, this disease is confined exclusively to English sheep. Mr. Isaac Seaman, in his prize essay on the subject, says:—"Parturient fever may be defined a disease of low, inflammatory character, involving, more or less extensively, the organs of reproduction, digestion, and respiration; the brain and spinal marrow are also involved. There is generally a greater determination of blood to some organs than to others; mostly, the uterus is first and principally affected; in some, the bowels and lining membrane of the abdomen (peritoneum); in others, the lungs; the brain and spinal marrow are often very much affected. It shows itself generally during the last twenty days' gestation, and within the first six days after parturition. The average duration of the disease is from seven to fourteen days; some die in two days, while others linger a month."

Mr. Seaman's essay is much too long for reproduction here, and we must content ourselves by simply quoting the above. The symptoms he mentions are numerous, and the treatment of a character unsuited to Australian sheep-owners. He mentions that pairs of lambs are frequently expelled—during the advanced stages of the malady—in a state of putrefaction; and that ewes that recover suffer afterwards from great weakness, and lose a great part of their wool.

CYSTITIS, OR INFLAMMATION OF THE BLADDER.—This, Mr. Spooner informs us, is a rather rare disease among sheep; and we do not deem it necessary to enter into any description of it.

SMALL-POX. — Although resembling the small-pox of men, this is a distinct disease, not communicable, either by contagion or inoculation, to men or children, or even to dogs or goats. Although common on the Continent of Europe, it was unknown in Great Britain for at least a century, until, in 1847, it appeared in Norfolk and the eastern counties of England, and, in the summer of 1862, in Wiltshire. Variolous sheep, or infected skins, appear in both cases to have imported the disease from abroad.

About ten days after exposure to contagion, the infected sheep become feverish, have a muco-purulent nasal discharge, and a hot, tender skin. Promptly and carefully must the sick be separated from the sound sheep; but if the spread of the disorder be not thus checked, inoculation may be resorted to. This will produce a mild form of the disease, and occasion a loss of from two to five per cent.; but, without these remedial efforts, the loss is likely to vary from 25 to 90 per cent.

Mr. Randall says, on this subject:—"*La clavelée*, as it is termed in France, attacks sheep at all seasons of the year, and in all conditions, but lambs sooner than grown sheep. Half, and not unfrequently two-thirds of a flock, used to perish by it. The sheep which recovers does not contract it the second time. It is communicated by contagion, and in every possible indirect way in which contagion is communicated among human beings, by substances which have been in contact with the subjects of the disease. A flock take it on being turned on a pasture which was occupied two or three months before by diseased animals, or by being driven over a road recently travelled by them." Mr. Youatt thus condenses and translates the statements of various French writers on this subject:—"In the regular *clavelée* there were four distinct periods; first, the symptoms which preceded the eruption, as dulness, loss of appetite and strength, and debility, marked by a peculiar staggering gait, the suspension of animation, and slight symptoms of fever. This continued during about four days, when commenced the second period, or that of eruption. Little spots of a violet colour appeared in various parts, and from their centres there sprung pustules, accompanied by more or less inflammation, isolated or confluent, and with a white head; their base was well marked and distinct; they were surrounded by a red aureola, and their centres were flattened. They were larger than an ordinary lentil. In some animals they were confined to a few spots, in others they spread over the whole body. They were scattered and thin, or disposed in the form of beads, or congregated together in a mass.

" When the disease was not of an acute character, and the eruption was not considerable, and the febrile symptoms were mitigated as soon as the pustule was developed, there was not much to fear: the eruption ran through its several stages, and no serious disorganization remained. But in too many cases the whole of the integuments became reddened and inflamed, the flanks heaved, the pulse, whether strong or obscure, increased in frequency, the mouth was hot, the conjunctive red, the breath fetid, the head swelled, the eyelids almost closed; rumination had ceased, the muscular power was exhausted, the pustules died away, with little apparent fluid secretion, a fetid diarrhœa ensued, and death speedily took place.

" The progress of the eruptive stage of the disease was frequently, however, a very unsatisfactory one. When the pustule had risen, and the suppuration had commenced, a new state of febrile excitement ensued, accompanied by more than usual debility. It lasted from three to four days, and during its continuance the pustules became whiter at their summit, and the fluid which they contained was of a serous character, yellow or red, transparent or viscid, and by degrees it thickened and became opaque, and then puriform; and at this period, when danger was to be apprehended, a defluxion from the nose ensued, and swellings about the head, as already described.

"This was the contagious stage of the disease, and when it was too easily and fatally transmissible, by accidental contact or by inoculation.

"Then came the last stage—that of desiccation; and, about the twelfth day from the commencement of the disease, the pustules subsided, or the integuments gave way, and the fluid which they contained escaped, and a scab was formed, of greater or less size and density, yellow or black, and which detached itself bodily, or crumbled away in minute particles, or powder.

"The contagion was now at an end, and the animal recovered his appetite and spirits, and strength. This stage of disquamation frequently lasted three weeks or a month.

"A secondary eruption occasionally followed, of an erysipelatous character. There were no distinct suppurating pustules; but there was a more serous or watery secretion, which soon dried.

"This was the regular and the fortunate course of the disease, but too frequently there was a fatal irregularity about it. Almost at the commencement there was excessive fever, and prostration of strength, and fetid breath, and detatchment of large patches of wool, and more rapid or bounding or inappreciable pulse, and strange swellings about the throat and head, and difficult deglutition. There was also a discharge of adhesive spumy fluid from the mouth, and of ichorous, or thick and yellow, or bloody and fetid discharges from the nostrils, often completely occupying and obstructing them. The respiration became not only laborious, but every act of it could be heard at a considerable distance; there was a distressing cough; the lips, the nostrils, the eyelids, the head, and every limb became swelled; the pustules ran together, and formed large masses over the face and the articulations; diarrhœa, that bade defiance to every medicine, ensued, and the end was not far off."

The symptoms of the disease after it appeared in England are thus described by Mr. Thos. Wells, of Norwich, in the *Norwich Mercury*:—"The leading symptoms of small-pox are a separation of the infected animal from the flock, a peculiar arching of the back, a drooping of the ears, a closing of the eyelids, amounting in some cases almost to blindness, and a pustular eruption, extending, more or less, over all parts of the body, but particularly those destitute of wool, or covered with hair only; such, for instance, as the cheeks, the skin under the arms and thighs, the under surface of the tail, udder, &c."

After the progress of the disease appears to be stayed, or after the application of remedial substances or medicines, it sometimes happens that the blood is not entirely freed from pock matter, in which case, Mr. Randall tells us, in an article which he quotes from Professor A. Numann, of Utrecht, that "it often produces (when the pox is already cured) a swelling in one or the other part of the body; as soon as such swelling is come to maturity it ought to be opened and the matter quickly removed from it."

Regarding inoculation, we may mention that a large body of sheep were so treated by various French veterinarians between the years 1805 and 1848, and that the percentage of deaths resulting from the operation was only one. It is also worthy of notice that sheep which have received the disease by inoculation have been re-inoculated and made to cohabit with diseased animals in the various stages of the malady, and it has been found that small-pox cannot be contracted a second time.

It is by no means impossible that at some future date, be it early or distant, small-pox may be introduced into the colonies. It is true that the *average* length of incubation is ten days. That is to say, that *about* that period of time will elapse between the time of the contact of a diseased with a sound and healthy animal; but it is not known what the *longest* time of incubation may be, and it is possible that it may be of sufficient duration to admit of the lengthy sea voyage to the colonies.

We have had more than one very narrow escape from the introduction of scab; indeed, but for the indefatigable scrutiny of the chief inspector of stock in New South Wales, the disease would undoubtedly somewhat recently have obtained a footing in that colony, from which its spread to the others would be merely a matter of time, and it is impossible to say to what extent its ravages might not have extended.

Should the colonies ever be so unfortunate as to suffer from the ravages of small-pox, we wish to impress upon all the necessity of inoculation. Immediately upon the discovery of the disease in any flock the sheep should be inoculated, and care should be taken that sheep which have escaped the contagion be not grazed upon land upon which the infected sheep have been depastured, until a considerable time has elapsed after the eradication of the disease.

SCROFULA.—Mr. Spooner thus speaks of this disease, which is one we have never seen any instance of :— " Sheep are liable to a scrofulous disease, which is almost uniformly fatal. It is called the *evil* in some places, and elsewhere receives other denominations.

" A hard swelling of the glands under the jaws is first observed ; after a time small pustules appear about the head and neck, which break, discharging a white matter, then heal, and are followed by others, more numerous. This gradually robs the animal of flesh, and, slowly pining away, it becomes at length quite useless, and in this state is destroyed. It seldom attacks great numbers at a time, but selects generally a few individuals from a flock."

Mr. Spooner recommends sending the animals to the butcher while in a fit state.

ABSCESS.—"This," says Mr. Spooner, "is a collection of pus, or matter, under the skin, and may be produced by a bruise, or by some constitutional cause. Whilst collecting, the surface of the skin is usually very tender, and sometimes there is also much constitutional irritation present. A collection of matter may be known by the heat, swelling, and pain of the part. On pressing it, the contained matter is felt to be fluid; and, the pressure being removed, the part immediately assumes its former shape, whilst a watery or dropsical swelling, on being pressed, leaves for some time the marks of the fingers. After some time the abscess points; that is, the matter can be more distinctly felt at one particular part, at which, if permitted, the abscess would soon burst. This, however, should not be permitted; but at this stage the abscess should be opened at the lowest part, or that which would admit most readily of it discharging itself."

This ailment is by no means uncommon in the colonies, and lambs, in particular, appear to be liable to it. We have always regarded it as constitutional, and believe it frequently arises from a deteriorated, unhealthy, or weak state of the animal's blood and system generally, which may arise either from insufficient nourishment or from some constitutional weakness inherited from either parent.

It is common to find lambs, at lamb-marking time, thus afflicted, and should the abscess have reached that stage of ripeness at which it is wise to open it, this should be done, but if not the animal had better be let go, and nature trusted to effect the cure, or, if the lamb be very weakly, destroyed, it not being generally convenient to await the further progress of the abscess.

BLINDNESS IN SHEEP.—This is sometimes, indeed frequently, caused by the grass seeds, which in some parts of the colonies are a terrible annoyance. They prevail during the summer months, and cause great destruction to the wool, as well as rendering themselves exceedingly troublesome to the bodies of the sheep. These they penetrate so thoroughly that it is not uncommon to find the livers of sheep invested with them ; and, in some places, at shearing time, some parts of the skin, especially the neck, arms, and brisket, present more the appearance of short-mown stubble than of the cuticle.

The presence of seeds in the eye may be detected, at first, by a running, which is followed by a thin white scum, and if, at or before this stage, the cause of disturbance be not removed, this will gradually thicken until the pupil of the eye is completely covered and total blindness ensues.

Care should therefore be taken, when the sheep are in the yards for drafting, lamb-marking, or any other purpose, that no sheep be allowed to pass without inspection, and, if any of the symptoms of blindness be apparent, examination of the eye.

A New Zealand correspondent of the *Australasian,* writing to that paper some time ago, says :—" Regarding blindness in sheep, I may state that I have had some experience in this disease. Some years ago I purchased a number of sheep from a northern province here (about 1,000), when, after being landed, changeable weather came on—warm days and cold nights, with snow and rain. I found that nearly half of them became quite blind, and were rushing about in all directions upon coming into contact with anything. I collected as many as I could in a yard, and bled them under the eye—the blood, I may tell you, was almost black in colour—and those I bled soon recovered their sight. The others that could

10

not be got in died.   The disease went through the whole flock to a greater or lesser extent."

In this case blindness arose, apparently, from great and severe, as well as unusual exposure, which probably caused a determination of blood to the brain.   The treatment exercised was quite in accordance with recognized authorities on this subject.

RABIES. — Among Australian flocks this disease is perhaps much commoner than is generally supposed.   It is certain that our sheep suffer greatly from the attacks and bites of dogs ; but, from what we can gather, it appears unlikely that dogs other than those in a domesticated state ever become rabid ; and this being the case, sheep liable to the attacks of real dingoes only are probably free from this disease.

From Mr. H. S. Randall's "Practical Shepherd" we select the following :—

" On Christmas Eve, 1862, some sheep belonging to my son, Henry P. Randall, were bitten by a dog.   I saw them next morning.   The flock consisted of about one hundred ewes, three years old (the previous spring), and in lamb.   I thought a dozen or more were wounded ; but as their hurts did not appear dangerous, I did not go to the trouble of ascertaining the precise number.   All were bitten, so far as I could discover, only about the head, and principally about the nose and ears.   The ears of some of them were torn into shreds, and their noses and lips covered with tooth-marks, showing that the attack on them had been long persisted in.   This was evidently the work of an animal which was unable to kill the sheep outright. The wounds on H. P. Randall's sheep were found to heal rapidly, and nothing was done to them. On the 12th January, 1863, he informed me that he had found one of the bitten sheep on the ground, unable to rise; that, on his helping it up, it moved about with difficulty. It had frothy saliva about its mouth. The next day it died. He had observed some of the ewes riding each other about, prior to the 12th, but did not know whether the dead one was one of these.

"On the 14th of January he informed me that two or three of the wounded sheep were riding and fighting each other; that one of them had suddenly butted him from behind; that, on his turning and offering to kick it, it would not retreat.

"I saw the flock in the afternoon; it was in fine condition. The wounds of the bitten sheep were mostly healed, and, with two exceptions, they looked as healthy and as full as any of the flock. Two of the sheep were obviously labouring under an attack of rabies. I continued to visit these and the succeeding cases daily, and generally twice a day, until the 29th of January, and until all the earlier cases observed by me (seven) terminated in death. I usually remained from three-quarters of an hour to an hour at each visit, carefully noting the appearance and actions of the sheep, and keeping a separate and continuous record of each case, as I was able to do without the least danger of mistaking one animal for another, as every one exhibited its number clearly printed on its side."

A complete history of this and another case was published in the annual volume of " Transactions of the New York State Agricultural Society," but being rather long, we content ourselves with a review of Mr. Randall's recapitulation of the history :—

"The cases I have described present variations in the minor developments of rabies, owing, perhaps, to individual peculiarities of the different animals ; but, as a whole, there has been a remarkable identity in the general symptoms.

"The first observed symptom in every case, which was seen at or near its commencement, was the same— viz., ungovernable apparent salacity (lust)—manifested, not according to the sex of the patient, all of which were ewes, and supposed to be in lamb, but in the manner in which the ram exhibits sexual heat. This resemblance extended to the minutest particulars in movement, postures, and in that characteristic note with which the male animal expresses desire as he approaches and importunes the female. The ewe incessantly attempted to ride her companions, but uniformly manifested rage, and turned and fought the one attempting to ride her.

"In all cases rumination was totally suspended from the first visible attack of the disease until death. The sheep manifested a depraved appetite, refusing the choicest preparations, and frequently eating wool from each other, and woodwork belonging to their pens. One was seen to eat the dung of another; another, snow which had just been saturated with urine ; and two eagerly to lick the mucus and saliva from the nose and mouth of a dead one. It was impossible to discover whether the rabid animals masticated their unnatural food or not.

" No exhibition of thirst was observed in any case, and, on the other hand, no dread of water when it was placed before them. Great and unnatural ferocity was exhibited by all the patients, and those which exhibited the greatest decrease of aggressiveness, as their strength failed, never resumed the usual timid habits of their nature. They retreated from nothing ; and to the last, if a man entered their pen and threatened them with a stick, they instantly attacked him. Some became blind, others partially so. The last day or two of their lives they staggered in their gait, fell over their dead companions, and rose with difficulty. Their debility was extreme. Even at this stage, and until actually dying, they manifested no stupor or insensibility to what passed about them, nor did they show the least symptoms of becoming paralytic.

" No remedies were administered to any of the sheep, under the impression that it would be utterly useless, and attended with disagreeable if not dangerous consequences."

We regret that considerations of space have compelled us to greatly reduce Mr. Randall's able report, but we have, in the above, placed before the reader the principal characteristics of this dreadful malady.

Mr. Randall deserves the thanks of all sheep-owners for the very careful and painstaking manner in which he has conducted this investigation, and his clear and lucid style of expression cannot fail to be admired by those who have had the opportunity of perusing his works.

There are scores of ailments to which sheep are liable of which we have not found space to make mention, having endeavoured to choose those for special consideration to the attacks of which Australian sheep are most likely open, and for the cure of which the means are sufficiently simple for the generality of Australian flock-masters—not requiring that manipulation of each sheep which the smallness of English flocks renders practicable.

The head of the sheep is subject to many maladies; amongst others, diseases of the horns, swelled head, sore face, swelled lips, and inflammation of the eye, but these we do not consider it necessary to describe, and will instead recommend those desirous of ascertaining full particulars on these points to consult either Mr. Youatt or Mr. Randall.

CHAPTER VIII.

SHEEP ON SMALL HOLDINGS.

MANY people erroneously suppose that to enter into the successful pursuit of sheep-husbandry a large area of country, backed by a large capital, is absolutely necessary, and that it is impossible on a small scale to depasture sheep profitably. And there are those who imagine that a large scope of country, and a large credit balance at their banker's, are all that is requisite for the success of their enterprise, and who at the end of a few years generally find that their lack of experience has greatly tended to reduce the aforesaid balance, if not to place it upon the other side of the ledger. We maintain that sheep-farming can be carried on as successfully and profitably, if not more so—in proportion, of course—upon 640 acres as upon 640,000 acres, and many can prove this who are now plodding along with a few milking cows, endeavouring to keep up a dairy upon land which has been exhausted by continual cropping. These farmers, if they try sheep-farming, can soon be established in neat homes, and with comfortable incomes, derived from land from which they would otherwise barely eke out a miserable existence; while at the same time they will be rendering their land more valuable year by year, and they will ultimately find their once worn-out soil almost as rich and productive as in its virgin state.

Of course, land, after being worn out by a continuation of crops, requires to be sown down with artificial grasses before the farmer can expect it to carry many sheep. For this purpose there are many varieties of grass, from which must be chosen that which is most suitable to any particular climate. Rye-grass and lucerne mixed together, say two-thirds of rye-grass to one-third of lucerne, make a very good pasture, and one that is more suitable to general climates than any other we know of.

Where these small holdings exist they are generally in close proximity to a market, where the produce of the farm, be it wool or mutton, can be disposed of at a trifling expense compared with that of the squatter, who has to transport his wool and stock many hundreds of miles before reaching a market. For this reason the proportionate profits of a small holder are greatly in advance of those of a large holder.

The most profitable sheep for the small holder to breed is that which yields the *most* wool and the *most* mutton. He should altogether disregard fineness of wool as a desirable quality in his sheep, because fineness of wool is quite incompatible with weight of carcass, which latter forms the principal element of small sheep-farming. The breeds of sheep we would therefore recommend are the Lincoln or Leicester, or a cross between either of these and the merino. By either of these crosses the value of the wool will be increased, and this will be accompanied by but little decrease in the weight of carcass; while the progeny are a more hardy and more easily fattened grade of sheep, they will consume less grass, and the lambs will always command a good price in the local market. One drawback to either of the above breeds— that is, the Lincoln or Leicester—is that they require a better description of fencing than the merino or many other breeds.

CHAPTER IX.

LAMB-MARKING.

SUCCESSFUL lamb-marking is one of the most important items in the management of a station; therefore, great care should be taken both in the cutting of the tails and in the castrating of the ram lambs. There are two methods of opening the scrotum, both of which have had many advocates. The method now generally employed is that of slitting the scrotum, not only because the operation is performed with more expedition, but because the cut heals more quickly, and is less liable to the attack of flies; and when the sheep goes to market the process has not detracted from the size of the scrotum, or purse. Great care must be taken not to cut too deep, but only sufficiently to allow the testicles to protrude far enough to be grasped by the teeth. The testicles should be drawn as cleanly as possible, and without any biting or breaking; and both should be drawn together.

Most people attach but little importance to the length the tails are cut, and on many stations but little regularity is observed. It is, however, a matter worthy of considerable attention, as it adds greatly to the appearance of a flock of sheep to see the tails of a uniform length, say from two to three inches; and a tail of that length, as well as the slitting of the purse, assists the buyer greatly in the judgment of a fat sheep. In a hot climate it sometimes bcomes advisable to dress the tail with tar, to repel the attack of flies. Stockholm tar, where procurable, is preferable, as it not only keeps off the flies, but assists in the healing; where it is not procurable, however, a mixture of kerosene and boiled oil, in the proportion of one-third of the former to two-thirds of the latter, will be found a good and inexpensive substitute. In no case should coal tar be used.

In the case of tailing well-bred ram lambs which are intended for stud purposes, it will be found advisable to attach a ligature about an inch from the end of the tail, so as to prevent excessive bleeding. The ligature should be taken off after the lapse of a few days.

The other method above alluded to is one worthy of but little comment; and the difference between it and the one we have described consists in the cutting off of the end of the scrotum, instead of slitting it. For many reasons, some of which we have already touched upon, this custom is rapidly falling into disuse; and we hope ere long to see it entirely abolished, as its many defects render it a very objectionable practice.

In docking, the tail should not be pulled too strongly, as, when that is done, the result is half-an-inch of bone left protruding beyond the skin, to heal slowly and cause unnecessary pain. Lambs should not be castrated while heated from over-driving; neither should they be castrated too late in the evening, especially if the nights are cold and frosty.

During a long and varied experience among sheep and stations, it has been a subject of considerable surprise to us to observe the little amount of judgment displayed and the evident lack of experience evinced by many managers, and even owners, in the manner in which lamb-marking has been conducted, especially on large stations. It will not, therefore, be out of place to offer a few suggestions regarding the arrangement of the yards and the working of the sheep, which we can best do by supposing a large flock of ewes and lambs ready in their paddock for cutting.

In ordinary seasons lamb-marking commences in June-

July. We will here suppose that the required number of lamb-catchers have been employed and despatched, under the overseer, with from 50 to 60 hurdles, for the purpose of arranging the yards in the paddock with which it is intended to commence. In most paddocks there is an old brush-yard; if not, a temporary one should be built if the distance is too far from the station drafting-yards. Having arrived here with the hurdles and stakes, they proceed to construct a lane on the opposite side of the yard to the gate; this should be about 9 feet wide, and as long as the number of hurdles will allow, spaces being left at regular intervals to allow the introduction of cross hurdles, so that the lane, when full, can be divided into small yards, thereby facilitating the catching of the lambs and lessening the danger of smothering. A small forcing-yard must be formed at the mouth of the lane by fencing off a corner of the yard. This will greatly assist the yarding into the race, and save the sheep much knocking about.

While this is being done, other men, despatched in the morning, have been engaged mustering the paddock, and by the time the yards are finished will probably have arrived with the sheep—about one-half the number the paddock contains—which are yarded to be in readiness for an early start next morning.

Supposing there are two fast cutters at work, they will require eight or ten catchers, and will probably cut from 5,000 to 6,000 lambs per day, if kept supplied with sheep. Before commencing, let a few grown sheep be thrown out for the purpose of steadying the lambs unti the first yard-full is cut and the ewes counted out. These sheep are then shepherded until the flock is finished and the musterers have arrived with the remainder; they are then let go, and those remaining are treated in the same manner.

SHEEP EAR-MARKS.

For the information of sheep-owners and managers we publish the following regulations issued by the Government of New South Wales:—

"OWNERS' TATTOO-MARKS.—1. All tattoo-marks used as owners' sheep-marks shall consist of not more nor less than three letters, or two letters and the sign &.

"2. The letters and signs used in owners' tattoo sheep-marks shall be of the Roman or script style or character, and shall be not less than $\frac{3}{8}$ of an inch in length and $\frac{1}{4}$ of an inch in width.

"3. All recorded owners' tattoo sheep-marks shall be marked in a clear and legible manner, as follows, namely:—

(1). Every such mark shall be imprinted on one or other of the following portions, and in the consecutive order in which they are here given, namely:—

First portion—The near or left ear.
Second portion—The off or right ear.
Third portion—The under side of the tail.
Fourth portion—Under the near fore arm.

(2). The breeder, or person imprinting the first recorded tattoo-mark upon any sheep, may do so on any of the portions hereinbefore mentioned.

(3). If the breeder's, or first recorded tattoo-mark, be made upon any portion other than the first, then the first portion shall in that case be held to follow the fourth portion.

(4). Every second or subsequent recorded tattoo-mark shall be imprinted on the portion which, according to the order hereinbefore prescribed, is next to that on which the immediately preceding mark is marked.

" 4. All sheep shall be deemed to be marked with the recorded tattoo-mark which shall appear to be the last mark imprinted upon such sheep, according to the order hereinbefore prescribed.

" 5. Notwithstanding anything contained in these regulations, the owner of pedigree sheep may, with the sanction of the directors, number such sheep with a tattoo-mark on the off ear, for stud purposes, and in that case the third portion shall be held to be the next in order to the first.

" 6. All applications for tattoo-marks shall, in terms of section 34 of the *Diseases in Sheep Act*, be made in the first instance to the inspector for the district in which it is intended to use such marks, who will make the necessary entries in his record, and forward the applications to the office of the Registrar of Brands, Sydney, to be recorded for the colony.

" 7. The fee for recording each tattoo-mark, including publication in the *Gazette*, will be six shillings (6s.)

" OWNERS' CUT EAR-MARKS.—8. The size of the cutters in the pliers for marking the undermentioned owners' cut ear-marks, shall not exceed the following :—

	Length.	Width.
" The Bayonet	1 inch	$\frac{1}{2}$ inch.
„ Club	$\frac{7}{8}$ „	$\frac{5}{8}$ „
„ Fork	$\frac{3}{4}$ „	$\frac{1}{2}$ „
„ Half halfpenny ...	$\frac{3}{4}$ „	$\frac{3}{8}$ „
„ Hole	$\frac{3}{4}$ „	$\frac{3}{4}$ „
„ Note	1 „	$\frac{1}{2}$ „
„ Slash	$1\frac{1}{4}$ „	$\frac{1}{16}$ „
„ Slit	$\frac{7}{8}$ „	$\frac{1}{16}$ „
„ Swallow-tail	$\frac{5}{8}$ „	$\frac{3}{4}$ „
„ Triangle	$\frac{5}{8}$ „	$\frac{5}{8}$ „
„ W (a double swallow)	$\frac{3}{4}$ „	$\frac{3}{4}$ „

" E. A. BAKER."

The *sex* is denoted by marking the owner's mark on the near (the left) ear of the male sheep, and the off (the right) ear of the female sheep, while the *age* mark must be made on the other or full ear.

"THE TATTOO-MARK.—Sheep-breeders in Germany, and we believe also in America, have long used the tattoo-mark for marking, or, rather, numbering, their sheep; and a few of the breeders in these colonies, and among others Mr. M'Caush, and afterwards Mr. Campbell, of Cunningham's Plains, have followed their example.

"Recently, however, Mr. Campbell made the suggestion that the tattoo should be used as a mark of ownership; and Mr. Bruce, the Chief Inspector of Stock, has worked out the idea, and devised a system of tattoo-marks, which will render Mr Campbell's suggestion of immense benefit to the colonies; for, used in conjunction with the 'cut' ear-mark, the tattoo will place the ownership of sheep on a far better footing than that of oxen, horses, and cattle under the *Brands Act*, enabling, as it will do, the rightful owner to substantiate his claim to his sheep when occasion requires, which the old loose system of ear-marking and fire-branding very seldom did. Under that system *ten* or *twenty* owners in the same district had the same 'cut' ear-mark, and although no two owners in the one district could use the same fire-brand, owners in different districts—though only, perhaps, a few miles apart—could do so, and all the best fire-brands were used forty or fifty times over in the colony. Then, again, as it is scarcely possible to fire-brand sheep without blotching, the thief could, without the risk of its being brought home to him, blotch or 'faik' a brand as it suited his purpose. The result of all this has been that convictions for sheep-stealing have seldom been obtained, and that crime is prevalent throughout the length and breadth of the colony.

"With the tattoo, however, none of these defects arise. In the first place, that mark *is given for the whole of Australia*, and no two owners in this or in any of the other colonies will have the same tattoo-mark; and in the next place, the owner in putting on a tattoo-mark will never blotch it, so that if there is any disfigurement in that mark it must be done wilfully, and the law is that if any person is found with a sheep having a wilfully disfigured mark on it, he will be liable to a penalty not exceeding £100. If, again, the thief were to cut the sheep's ear off, that renders him liable to the same penalty, and he would not, therefore, dare to cut the ear with the tattoo-mark off.

"Hitherto the tattoo-mark has been made on the living sheep's ear by snapping the ear with the tattoo ear pliers, or stamping it with the tattoo stamp, and then rubbing in the Indian ink; and, marking in this way, from 1,800 to 2,000 grown sheep can be marked in a day. It is to be hoped, however, that a more speedy mode of using the tattoo will be found. It answers very well on the ear of the dead sheep to snap a pad saturated with ink with the pliers, and then snap the ear—like bringing the stamp on to an inked pad and then stamping the letter, in post-office work—and it is believed that means will be found to make this mode of tattooing answer equally well on the live ear, when the marking would be done in less than half the time it now takes in rubbing in the ink."—*From the Town and Country Journal.*

The plan described above for tattooing answers excellently on live sheep. We have tried it on lambs, and it showed quite distinctly two years after the tattooing took place. One man can tattoo from 2,500 to 3,000 per day, the tattoo-mark being made at the same time as the lambs are castrated.

APPLICATION TO RECORD A SHEEP BRAND OR MARK IN NEW SOUTH WALES.

To

Inspector of Sheep.

Owner's Brands.   I have to request that you will record the brand and marks on the margin hereof as the

Owner's Mark.	sheep brands and marks to be used on station of , of which is the post town, and on which there are now sheep and lambs belonging to
Age Mark.	and that you will also enter in your record that the age of the sheep on the station is noted by and the class by
Class Mark.	I enclose the authorized fees, amounting to £ , particulars of which are given in the schedule below.

Witness—

Owner or Super.

Particulars Recorded.	Number of Sheep.	FEES.	
		Rate each.	Amount.
Recording owner's brands			
,, ,, mark		5/	
,, distinctive age mark		1/	
,, ,, class mark		1/	
Publication in *Gazette* .... each ..		1/	
TOTAL			

WHEN RECEIVED.

Date.	Time.

CHAPTER X.

PADDOCKS *v.* SHEPHERDING.

PREVIOUS to the great gold rush of 1852 the system of shepherding was in general use, and the existence of paddocked sheep at that time was quite as rare as is that of shepherded ones now. Although we occasionally meet with an advocate of the old system, almost all our sheep-owners have adopted the new one, and in the course of a few years we may expect shepherding to exist but in memory.

On a run completely fenced and subdivided the stock enjoy perfect quiet and freedom from disturbance, except such as may be caused by the traffic on public roads; where it is practicable, however, it is advisable to place strong sheep in paddocks traversed by such roads, and also where the distances from water and the drafting yards are the greatest. The quietest and most secluded paddocks should always be chosen for the ewes during lambing, the reasons for which are obvious. When it is necessary to go through and among the sheep when lambing, it should always be done as quietly as possible, and in no instance should a dog be allowed in the paddock, as the presence of such a visitor among lambing ewes is almost certain to create a commotion, and cause the lambs to separate from their mothers. Excepting in the cases of maiden ewes and bad seasons, attention to such precautions as these, and a strict attention to the welfare of the sheep, will result in a percentage of lambs equal to that attained by the most careful of shepherds, although in exceptional cases we have known this to exceed 100. We may mention that a paddock for lambing ewes should in no case exceed 2½ miles square.

RAM PADDOCK.

The essential requirement of a ram paddock is security. The fences should not be made with wire when anything else is procurable, especially if it is intended to depasture ewes in the adjoining paddocks. We have seen rams placed in a paddock surrounded by a wire fence that for any other description of sheep would have been perfectly secure, the wires being well strained and the gauge of an ordinary sheep-proof description; in the two adjoining paddocks were ewes, and the rams appeared to experience no difficulty in joining them. By having the ram paddock enclosed by a good chock and log fence the evil effects of constant lambing will be avoided. Rams should never be without salt, either natural or artificial.

SHEEP LIVING WITHOUT WATER.

The *Lebanon* (Pennsylvania) *Courier* prints the following extract from a letter from Stehman Forney, of the United States Coast Survey, dated on the Island of San Clement, in the Pacific:—"I am at present engaged making a survey of San Clement Island. It is forty miles from the mainland, and is twenty-two miles in length and two miles wide. It is a wild, dreary place, with no water on it, except in immense natural tanks, which are so deep and precipitous that the water in them is inaccessible. I transport the water for my men and horses from the mainland. There is no wood, either, on the island, which is of volcanic formation, and composed of lava and conglomerate. The top of the island is covered with an abundance of grass, which sustains about 10,000 sheep, and strange to say, they live, grow very fat, and are very profitable to their owners, and yet in the summer season get no water except in the form of dew on the grass. There is, however, a peculiar plant on the island, called the ice-plant, which is filled with moisture, and is eaten by the sheep to quench their thirst. They are very fat, and make the finest mutton I have ever seen." Sheep will not live on dry feed any length of time without water, even in the winter time.

CHAPTER XI.

GRASSES.

BENT GRASS.—The species are numerous, and are found in almost all countries and climates; several are natives of Britain. All of them are grasses of a slender and delicate appearance. Some are very useful as pasture-grasses and for hay, upon account of their adaptation to certain kinds of soil, although none of them are regarded as very nutritious. The *Common Bent Grass* forms a principal part of the pasture in almost all the elevated districts of Britain, and is equally abundant in many parts of the Continent of Europe. It resists drought better than almost any other grass, but is only sown by agriculturists on soil unsuitable for the more luxuriant grasses. In light, dry, cultivated grounds, it is often a troublesome weed, known as black squitch, or quick-grass, and frequent harrowing is resorted to for the removal of its creeping perennial roots. It is as frequent on wet as on dry soils, and varies much in size and appearance. The *Marsh Bent Grass* is also very common in Britain, and forms a large part of the natural pasture in many moist situations, and is very similar to the species just described, but generally taller and stouter.

FIORIN GRASS is a useful grass in moist grounds, newly-reclaimed bogs, or land liable to inundation. It was unduly lauded, and the consequent disappointment led to its being unduly disparaged. The first three joints of the culms lie flat on the damp soil, emitting roots in abundance, and it was formerly propagated by chopping these into pieces and scattering them, but now generally by seed.

HERD GRASS is a native of the United States, with broader leaves than either of the preceding species, very creeping roots, and large pinnacles almost level at top. It was at one time strongly recommended for cultivation, but has gone out of repute in Britain. It is, however, more highly esteemed in France, particularly on account of the great crop which it yields on deep sand and on moist calcareous soils.

BROWN BENT GRASS.—A common perennial British grass, abundant in moist heaths and moorish grounds. Is valuable for mixing with other grasses to form permanent pasture on poor, wet, peaty soils.

SILKY BENT GRASS is a beautiful grass, with very slender branches to its ample pinnacle, which, as it waves in the wind, has a glossy and silky appearance. It is a rare native of sandy grounds in England, common in southern and central Europe; an annual grass, occasionally sown in spring to fill up blanks in grass-fields.

CLOVER, or TREFOIL, contains a great number of species, natives chiefly of temperate climates, abounding most of all in Europe, and some of them very important in agriculture as offering pasture and fodder for cattle. The name clover is, indeed, popularly extended to many plants not included in this genus, but belonging to the same natural order, and agreeing with it in having the leaves formed of three leaflets, particularly in those of them which are cultivated for the same purposes, and sometimes collectively receive from farmers the very incorrect designation of *artificial grasses*, in contradistinction to *natural grasses, i.e.,* true grasses. The true clovers *(trifolium)* have herbaceous, not twining, stems; roundish heads, or oblong spikes of small flowers, the corolla remaining in a withered state till the ripening of the seed; the pod enclosed in the calyx, and containing one or two, rarely three or four seeds. About 17 species belong to the flora of Britain.

COMMON RED CLOVER is the most important of all to the British farmer, and is a native of Britain, and of most parts of Europe, growing in meadows and pastures. It is a perennial, but is generally treated as if it were a biennial. Its heads of flowers are oval or nearly globular, very compact, about an inch in diameter, purple, more rarely flesh-coloured or white.

ZIG-ZAG CLOVER, also called *Meadow Clover, Marl Grass* and *Cow Grass*, much resembles the common red clover, but is easily distinguished by the smooth tube of the calyx. The stems are also remarkably zig-zag. It is a common plant in Britain and most parts of Europe.

WHITE, or DUTCH CLOVER, is also a common native of Britain, and of most parts of Europe. When a barren heath is turned up with the spade or plough, white clover almost always appears. It is said to be a native also of North America, where, however, it is perhaps only naturalized. The flowers of all kinds of clover are the delight of bees, but those of this species perhaps particularly so.

ALSIKE CLOVER, a perennial, regarded as intermediate in appearance between the common red clover and the white clover, has of late attained a very high reputation.

CRIMSON, or ITALIAN CLOVER, an annual, native of the south of Europe, is much cultivated in France and Italy, and has of late been pretty extensively grown in some parts of England, producing a heavy crop.

MOLINER'S CLOVER very much resembles crimson clover, but is biennial, and has pale flowers. It is cultivated in France and Switzerland.

ALEXANDRIAN, or EGYPTIAN CLOVER, an annual species, a native of Egypt, universally cultivated in its native country, where it is the principal fodder for cattle, has been tried in Britain, but the colder climate has been found to render it less luxuriant and productive. It is supposed to be one of the best kinds of clover for the British colonies. It has oval heads of pale yellow or whitish flowers.

YELLOW CLOVER, or HOP TREFOIL, is very common in gravelly soils in Britain, but not much esteemed. It has smaller leaves and heads of flowers than any of the cultivated species. Its flowers are yellow.

It is little more than a century since clovers were introduced into field culture in Britain; they are now universally cultivated on large farms, in alternation with grain crops. The kinds most generally sown are the *Common Red, Cow Grass, Dutch White, Yellow,* and *Alsike.* The *Common Red* is the finest and most valuable, but it is difficult to grow unless on naturally rich soils. In America it grows well on sandy loams, though sown every alternate year on the same land. But in Britain the land is thought to become " clover sick " when sown too frequently with this crop. An interval of not less than eight years is thought advisable. From 6 lbs. to 20 lbs. of seed per acre is the quantity sown. Red clover is most esteemed for being mixed with rye grass for the making of hay. When it grows well, it bears to be cut more than once a year. *Cow Grass* much resembles the common red clover. It is coarser, but hardier, and better suited for pasture, as it bears more herbage, and comes better up after being eaten close down by stock. *Dutch White Clover* is only esteemed for pasture; it grows short and thick on the ground, but throws out fresh stems and flowers during the most of the growing season. In the south of England it is sometimes sown with but little rye grass seed along with it; in Scotland as much as a bushel and a half of rye grass is mixed with it for pasture. *Yellow Clover* is chiefly sown on ground where neither the white nor red grows freely. It is not sown so frequently as it probably ought to be; for it rises early in spring, and a mixture of it with other clovers forms good pasture on all grounds. *Alsike Clover* rises much higher than white clover, and offers to be a useful addition to our pasture plants. Land must be thoroughly cleaned of perennial weeds before it is sown with clover, as the land cannot be subjected to cultivation while it is under this plant; clover, therefore, is always sown in the end of the rotation, or as near the fallow or turnip crop as possible. It is sown early in spring, and slightly harrowed in; for the seeds, being small, are not difficult to bury. Farmyard manure is as good as any for clovers. A well-manured soil greatly assists in keeping the plants from dying out in spring. Clovers, like grasses, play a most important part in restoring fertility to land which has been exhausted by grain crops. Their leaves gather food—carbonic acid and ammonia—from the atmosphere, which they store up in their roots and stems; and these, on decomposing, afford food for cereals, or other crops which are more dependent on a supply within the soil.

COTTON GRASS.—The species are not very numerous; they are natives of the colder regions of the northern hemisphere, and are said to be valuable for sheep pasture.

COUCH GRASS, also called *Wheat Grass, Dog Grass*, a grass which, although of the same genus with wheat, is chiefly known to British farmers as a troublesome weed. It is common in most parts of Europe and North America. It grows to a height of 1½ to 3 feet, and has two coned spikes and flat spikelets. It is perennial, and its creeping roots render it extremely difficult of extirpation; they are carefully gathered out of land under cultivation, but they make the plant very useful in fixing loose, sandy soils, so as to form pasture. It is not, however, esteemed a very nutritious grass.

CROWFOOT.—The species are numerous, mostly perennial. Some of them adorn meadows with thin yellow flowers, familiarly known as *buttercups*, others known by the name of *crowfoot*. Pastures in which they are very abundant are injured by them; they are particularly supposed to give un unpleasant taste to milk and butter; but it is thought not improbable that a moderate mixture of these plants with the other herbage is even advantageous, and that they may act as a condiment. Their acridity is lost in drying, and they are not injurious to hay.

RAPE, or COLESEED, a biennial plant, much cultivated, both on account of its herbage and its oil-producing seeds. It is a native of Europe, and, perhaps, of England; but it is hard to say where it is truly indigenous, and where naturalized. The root of the rape is slender, or in cultivation sometimes become carrot-shaped. The stem is taller than that of the turnip, and the foliage more luxuriant. Rape delights in a rich alluvial soil, and is particularly suitable for newly-reclaimed bogs and fens. In rich soils rape sometimes attains a height of 3, or even 4 feet, so that the sheep turned in are hidden under the leaves, and seem to eat their way into the field. They eat the stalks even more greedily than the leaves. A too exclusive feeding on rape is, however, apt to produce diseases, which a supply of salt is found useful in preventing.

RYE GRASS.—This grass is highly valued for forage and hay, and is more extensively sown for these uses than any other grass. It grows well even on poor soils.

ITALIAN RYE GRASS is a native of the South of Europe, and is much esteemed as a forage and hay grass. In many soils and situations it succeeds extremely well, and is remarkable for its verdure and luxuriance in early spring. It is preferred by cattle to the common rye grass. There are many varieties of rye grass. It is generally sown along with some kind of corn, and, vegetating for the first year amongst the corn, appears in the second year as the proper crop of the field.

LUCERNE.—One of the most valuable of the leguminous plants cultivated for the supply of green food to cattle. It is a native of the south of Europe, and has been cultivated there for an unknown antiquity. It is not very largely cultivated in Britain, although in some places very successfully; but the climate of Scotland is not too cold for it, and the different results obtained by farmers who have tried it seem to depend chiefly on differences of soil and management. It is largely cultivated in some parts of North and South America, and in Peru with great success, both on the coast, in all the heat of a tropical climate, and on the mountains, to a height of more than 11,000 feet above the level of the sea; flourishing, however, only during the moister part of the year in the former situation. It endures great droughts, its roots penetrating very deep into the ground, but loves a rich and calcareous soil, and never succeeds on damp soils or tenacious clays. It is a perennial, and affords good crops for a number of years. It may be mown several times a year, growing very quickly after being mown. The quantity of produce is very great, and no other forage plant is ready for use so early in spring. It ought to be mown before it comes into flower, as it then becomes more fibrous, and less succulent and nutritious.

REED CANARY GRASS is a somewhat reed-like grass, 4 to 6 feet high, with creeping roots, which help to secure river banks, &c. It yields a great bulk of hay, but has been very generally despised as a coarse grass fit only for littering cattle.

MEADOW GRASS.—The species are very numerous, chiefly natives of temperate and colder parts of the world, and forming in these a very important part of the herbage of pastures and meadows. The herbage is tender, nutritious, and rather abundant. The *rough-stalked* and the *smooth-stalked* are esteemed among the most valuable for sowing in mixtures of grasses for pasture.

The ABYSSINIAN MEADOW GRASS, an annual species, yields immense returns of herbage in its native country, but a warmer climate than that of Britain seems to be requisite for its cultivation. It is employed with advantage for sowing on greens in towns, and wherever, from any cause, perennial grasses are apt to be destroyed.

FESCUE.—A genus of grasses very nearly allied to brome grass. The species are numerous, and are very widely diffused over the world, both in the northern and southern hemispheres. Among them are many of the most valuable fodder and pasture grasses. None are more valuable than some of the British species. *Meadow Fescue*, a species from 2 to 3 feet high, common in moist meadows and pastures of rich soil, in Britain and throughout Europe, in Northern Asia, and in some parts of North America, is perhaps excelled by no meadow or pasture grass whatever. It is suitable for both alternate husbandry and for permanent pasture. *Hard Fescue*, a grass from 1½ to 2 feet high, is one of the best grasses for lawns and sheep-pastures, particularly on dry and sandy soils. *Creeping Fescue* is probably a mere variety of the preceding one, being distinguished chiefly by its extensively-creeping root, which particularly adapts it to sandy pastures and to places liable to occasional inundations. *Sheep's Fescue* is a smaller grass than any of these, not generally exceeding a foot in height, and often much less, abundant in mountainous pastures, and especially suitable for such situations, in which it often forms a principal part of the food of sheep for many months of the year. It is common in all the mountainous parts of Europe, and in the Himalayas ; it is also a native of North America, and species very similar, if not mere varieties, abound in the southern hemisphere. Its habit of growth is much tufted. *Tall Fescue* is a grass of very different appearance, 4 or 5 feet high, with spreading, much-branched pinnacle, growing chiefly near rivers and in moist, low grounds, and yielding a great quantity of coarse herbage, which, however, is relished by cattle. All these species are perennial.

BROME GRASS.—The species are numerous, and some of them are very common British grasses—none more so than the *Soft Brome Grass*, an annual or biennial, which has very soft, downy leaves, grows well on poor soils, and is readily eaten by cattle, but is not much esteemed by farmers, either for the quantity or quality of fodder which it yields. Its seeds have also the reputation of possessing deleterious or poisonous properties. *Rye Brome* is generally regarded as a troublesome weed, especially in fields of rye. In a young state it has a great resemblance to rye. Its seeds, which are large, retain their power of germination for years, and do not lose it by passing through the intestines of animals.

There are nearly 4,000 known species of grasses—about one-twentieth of all known phanerogamous plants. They are distributed over all parts of the world; some are characteristic of the warmest tropical regions, and some of the vicinity of perpetual snow; but they abound most of all—and particularly in their social character, clothing the ground with verdure, and forming the chief vegetation of meadows and pastures—in the northern temperate zone. There is no kind of soil which is not suitable to some or other of the grasses; and, whilst some are peculiar to dry and sterile soils, others are only found on rich soils with abundant moisture; some grow

in marshes, stagnant waters, or slow streams, some only
on the sea coast; none are truly marine.

CHAPTER XII.

SHEEP-DROVING.

DROVING, to be properly and economically performed,
requires, perhaps, as much experience and care as any
work among sheep. One of the principal faults of many
drovers is the length of stages which they travel. Some
slight extenuation may be accorded them, however, when
we consider the inconsiderate manner of but too many
squatters in urging them on beyond the limit fixed by
law, and in refusing them the means of supplying their
sheep with water. Drovers are also sometimes compelled
to lengthen their stages in order to reach water, even if
the squatters were willing to give or sell it. The Govern-
ments of the different colonies, and of New South Wales
in particular, have been slow in remedying this evil;
the latter have, however, commenced at last, by sinking
tanks on the One-Tree Plain, and other places between
Hay and Wilcannia which lie on the direct route to
Melbourne from Queensland. Another reason, but by no
means an excuse for long stages, is the payment by
squatters *per capita*, according to distance, for delivery,
which is a system we have no hesitation in condemning.

Fat sheep should not be travelled more than from six
to eight miles per day unless under very exceptional
circumstances, and we consider that under ordinarily

favourable circumstances ten or eleven miles per day is sufficiently far for any sheep to travel. We have known a manager foolish enough to allow his employer's sheep to be travelled at the average rate of eighteen miles per day, the longest stage being twenty-one miles, sixteen of them being covered before dinner-time, and this without any reasonable cause whatever. Such cases, however, happily prove the exception rather than the rule.

There are two methods of payment of drovers which we advocate, and would recommend in preference to the one we have mentioned: the first and best being the payment of so much per week, according to the number of sheep, and country to be passed through, the drover to find the plant—*i.e.*, horses, tents, &c.—and pay all wages and expenses. The other plan is to pay the drover at the rate of so much per week, generally from £3 to £6, and he is furnished with the means of paying all wages and expenses, such as crossing bridges, tolls, &c. In this case the squatter finds the plant. Either of these plans will be found more to the advantage of the sheep-owner than that of paying so much per head. He will have his work better done, and it need cost no more, if he is able to make a correct calculation of what the probable cost of droving will be.

If sheep have been without water for any length of time, they should never be allowed to go to drink in a large body, and in no case without a thorough inspection having been made of the place where they are to water. As an instance of the consequences of neglect of this precaution, we may mention the case of a drover who had to travel with 10,000 sheep a distance of 60 miles without water; at the end of this journey there were two large waterholes seldom seen dry. When distant a mile or so from these holes, the sheep, smelling the water, began to run, and were allowed to do so. They rushed on in a long "string," and when the drovers came up with the last ones they found the water had nearly all dried up, and left nothing but a quagmire, in which were smothered some 25 per cent. of the flock.

Sheep should not be travelled in flocks larger than from 1,500 to 2,000, especially fat sheep for market, because they have no chance of feeding if travelled in larger mobs. Even on stations sheep should not be driven in too large mobs.

Every person travelling with sheep requires a "Permit to Travel," and must also have a "Travelling Statement." The permit is obtained from the inspector of sheep for the district, and the statement from the owner or superintendent of the station from which the sheep start. We give below the forms of a "Permit to Travel" and of a "Travelling Statement" for New South Wales, which require to be filled in, and the former sent to the inspector of stock for his signature, and the latter kept in possession of the drover, and signed by the owner or superintendent.

PERMIT TO TRAVEL (N.S.W.)

This is to certify that the sheep more particularly referred to in the schedule below are hereby permitted.

to travel to their destination by the route specified in the said schedule.

Schedule referred to above.

Number.	Description.	Brands and Ear Marks.	Name and Address of Owner and of Person in charge.	From what District and Run.	Route permitted to travel.	Consignee and Destination.

188

Inspector.

TRAVELLING STATEMENT.

I,   of   , do solemnly declare that I am the owner (*or* superintendent of Mr.   , the owner) of the travelling stock more particularly described in the schedule below, and I further declare that the said stock are this day to be taken by me (*or* by   , as my drover) from (*state the name of place and run*), and are intended to be driven by me (*or* him) to   , being their destination (*state the name of place or run*) by the following route, which is an ordinary (*or* the direct) route, namely :—

Schedule referred to.

Number of Stock.	Description of Stock.	Sex.	Marks.	How and where Branded.	Diseased or Sound.

Owner (or Superintendent).

Signed at   this   day ⎱
of   188   ⎰
Witness—

CHAPTER XIII.

SHEEP-WASHING.

" THE utility of washing sheep before shearing is the subject of a good deal of discussion. One class of producers advocate it, on the ground that it prevents a useless transportation of dirt to market, that it improves the saleableness of wool, and that it avoids the operation of an unequal rule of shrinkage, applied by buyers indiscriminately to all unwashed wools. Another class of producers contend that it is injurious to the health of sheep, and that it subjects sheep to the danger of contracting contagious diseases; and, finally, that any custom of buying, or conventional rule of shrinkage, which is found unfair in itself or opposed to public utility, should be promptly abandoned. The objection of transporting dirt is a good one, unless it secures some advantage which counterbalances its cost. I am satisfied that washing, properly conducted, in water of suitable temperature, is not in the least injurious to decently hardy sheep; not any more so than an hour's rain any time within a month after shearing.

" And what sound objection can the *buyer* have to the farmer shearing his sheep unwashed, if he chooses to do so ? If the farmer sends *dirt* to the market, he—not the buyer—pays for the transportation. Washed or unwashed, the wool must go through the same cleansing process. Am I asked if the buyer has not the right to judge of the condition in which *he* shall voluntarily purchase a commodity with his own money ? By no sound principle, either of morals or commerce, have any class of buyers a right to establish rules of purchasing, not necessary to protect their own legitimate interests, which are calculated to injure the interests of producers.

" The rule that all wools shall be washed, or subjected to a deduction of one-third, to put them on a par with brook-washed wools, operates very unequally. A large, highly-yolked ram, housed in the summer, will have at least two pounds, and a ewe one pound, more yolk in its fleece than would the same animal if unhoused in the summer. Should the unwashed wool, then, sell at the same rate of shrinkage in both cases ? If we were to admit that one-third is a fair *average* rate of shrinkage on all unwashed wools, is there any justice in making the producer of the cleaner ones suffer for the benefit of the person who chooses to grow yolkier wools? Does the manufacturer wish to pay a premium on the production and preservation of yolk in wools?

"No manufacturer claims that the present rule of shrinkage operates strictly equitably in all cases; but some manufacturers contend that a discrimination in unwashed wools would be impracticable, or at least inconvenient, and that if the present rule injures the interest of the producer, all he has to do is to wash his wool. It would be difficult for anyone to show that there is any greater practical inconvenience in deciding between the different amounts of yolk in unwashed wool than there is in deciding between the different amounts of foul seed in wheat and other varieties of grain, of useless weeds in hay, or even of *yolk in washed wool;* yet who thinks of buying these impure commodities at a fixed rate of shrinkage? Still less excuse is there for preserving an arbitrary and unequal rule, as a quasi *punishment* on growers who only believe themselves consulting their own legitimate interests, and who certainly are not invading those of others.

"The only sound and equitable course is to abolish any fixed rule in the premises—to buy unwashed wool as wheat, other grain, hay, and washed wool containing impurities are now bought—viz., subject to a deduction proportioned to the amount of impurity in each particular case, *clean wool* being made the standard. It is as easy for the buyer and seller to agree on the amount of deduction as to agree on the quality.

"I have in this connection spoken only of the manufacturers as buyers, though, directly, other classes of buyers are equally concerned in the question. But I have done this on the supposition that, as all wools go ultimately into the hands of the former to be prepared for consumption, their action in the premises would be the controlling one among all classes of purchases."—*H. S. Randall.*

CHAPTER XIV.

SHEEP-SHEARING.

THE shearing of sheep is an olden custom—as old as the days of Nabal, who had three thousand sheep and a thousand goats, and who "sheared his sheep" in Carmel.

In sheep-shearing, the position in which the sheep is held constitutes the greater part of success; and when this difficulty is overcome, very little practice will enable almost anyone to remove the fleece with very fair success. In fact, the proper position of the sheep constitutes, to shearing, the same importance as does the selection of a well-bred ram to the production of good wool, and is a matter of equal importance to the squatter and shearer. To the squatter—by less injury to the sheep, which will, if properly held, escape many cuts which would otherwise be inflicted, and will also incur less danger of being internally injured—it is a matter of great importance; to the shearer, it not only gives less labour, but a better result in every respect.

The sheep, in the first instance, should be *carried* out of the pen, and carefully laid on the floor. If it is not intended to enforce this, or indeed any rule, it should be excluded from the agreement. Although it does not injure a sheep to be *carefully* taken out by the hind leg, shearers have a very injurious method of twisting the leg in throwing the sheep when out of the pen, and one that is very liable to either break the leg or dislocate either one or other of the joints. Therefore, every man should be compelled to carry his sheep from the pen, by placing one hand under the brisket and lifting the sheep bodily from the floor; this should be done even with grown wethers.

The belly-wool should, in all cases, be removed first; detached from the fleece, and thrown into the middle of the floor. The crutch should then be thoroughly cleaned, and all the trimmings taken off the inside of the hind legs; it is also advisable to have the wool taken from the outside of the near hind leg at the same time. The fleece should be opened up the neck, commencing at the brisket; in no case allow the fleece to be opened half-way up the neck, or, as some shearers do, under the lower jaw. While opening the fleece, *both* blades of the shears must be kept under the wool, and close to the skin; otherwise a wide strip of wool, sometimes three or four inches wide, will be cut in two, and rendered almost useless. In shearing the first side of the sheep, each blow should be continued round until the back-bone is passed ; this avoids the second cut caused by the blow up the back, which should not be allowed, as the " cutting through " which results considerably depreciates the value of the wool. This habit, we are sorry to see, is gaining favour among shearers every day, and is one that any sensible man will check at once. When the shearer has shorn the first side, and is in the act of shearing the tail, it should be closely observed that he does not press his knee or foot on the sheep, as it is at this time that the animal is most at the mercy of the shearer ; and the latter, if he be of a cruel or vindictive disposition, can, with less chance of being observed, easily inflict unnecessary and, in many instances, permanent pain to the animal. It is also very easy for a shearer to leave a considerable quantity of wool on the last side by not shearing round sufficiently far to cut what he left in taking off the belly-wool ; this, however, rarely occurs, except with either inexperienced men, or men who are attempting to work beyond their speed.

The person who has the supervision of a lot of shearers should refrain, as much as possible, from useless faultfinding ; when he has occasion to speak, let it be to the point, sharp and decisive. Continual fault-finding causes discontent among the men, and considerably reduces that *prestige* which all managers should possess. Correct a shearer once ; and if he requires correcting a second time, let that be the last : dismiss him at once. If a manager allows his men to see that he means what he says, he will get his work done far better and much more agreeably.

" Running," or " ringing," in shearers' phraseology, is a practice that should always be discouraged where it is desired to insist upon a good clean cut, as it increases the speed, and therefore decreases the quality, of the shearing, not only of the men who are trying to out-number each other, but has a deteriorating effect upon that of all the men employed.

Do not allow the shearer to tread upon, or tear, his fleece ; this a careless man is very liable to do, especially when turning out his sheep after shearing. If a man turns a sheep out without tarring the cuts, send him out to do so, or make him bring the sheep back on the board. Have any serious wounds sewn up and well tarred, as the wound heals quicker, and is not so liable to the attack of flies. Make every shearer let you know when he meets with a sheep whose horn is growing into its head or eye, and at once saw it off with a small saw kept in some handy place. The horns of the rams should be carefully looked to at shearing time, as they often grow into the sheep, and get broken while fighting. Do not allow the boys engaged in picking up the fleeces to do so until the sheep is turned off the board, as a great part of the fleece is sometimes held by the sheep resting on it, just prior to being turned out, and is con-sequently torn off. Have the board swept as soon as the shearer finishes shearing his sheep, and before he takes another from the pen. All the belly-wool must be picked up as soon as thrown aside by the shearer, and kept separate. Two or three baskets, placed in the centre of the floor, will be found useful for this purpose ;

these are emptied at every spell. Have the piece-wool divided into first and second pieces—all the stained pieces, or dags, going with the locks.

During shearing, care must be taken not to overcrowd the shed, as sheep are often injured, and sometimes smothered, through this. It is a great fault with many people, especially if the weather looks threatening.

BRANDING.—All the sheep should be carefully branded after shearing. Tar is generally used ; but common fat and lampblack answer equally well, and do not injure the wool so much. Avoid dogging the sheep, while in the yard, as much as possible, and let their time in the yard be as limited as possible.

Wool, when pressed, should be stored in a dry place. It is a very common, but erroneous idea, that wool gains in weight for a long period after pressing ; it does so for about twelve months ; but every wool-buyer knows that after that period it loses several per cent. by the evaporation of yolk and moisture.

SHEARER'S AGREEMENT.

I hereby agree to shear for Mr.   the sheep now depasturing on the   station, or such portion as I may be directed to shear, at the rate of   shillings per hundred sheep, finding my own rations. I also hereby agree to shear the sheep in a proper, workmanlike manner, and according to the following specifications :—

SPECIFICATIONS OF SHEARING.

1. Bellies to be taken off in one unbroken piece, and placed on one side.

2. In opening the fleece at the lower part of the neck, both blades of the shears to be kept under the wool, and close to the skin, so as to avoid twice cutting.
3. No after-cutting allowed on the body of the sheep.
4. No shearer allowed to put his foot or knee on any sheep.
5. In shearing lambs, the shearer to shear the belly and points first, and then move the lamb so as to avoid mixing the fleece with the belly-wool.
6. No shearer to catch sheep while his pen is being filled.

I also further agree to abide by the following rules :—

RULES OF THE SHED.

1. That no money shall be paid to any shearer, nor any amount paid on his behalf, till the completion of shearing.
2. That the manager of the shed shall have the power of dismissing any shearer who does not shear to his satisfaction, or who misconducts himself in the shed.
3. That obedience shall be given to all reasonable commands given by the manager, and that all regulations he may make as to the hours of shearing, and manner of shearing, are to be complied with.
4. That in the event of any shearer being dismissed during the shearing, such shearer shall be paid at the rate of   shillings per hundred sheep shorn, and pay for his board at the rate of   shillings per week.

On the completion of the shearing, all the rules and regulations contained in this agreement being complied with, the said   hereby agrees to pay the said   , shearer, at the rate of   shillings per hundred sheep shorn.

Rations to be supplied at the following rates :—

Flour,	per bag	Wethers,	each.
Tea,	per lb.	Ewes,	,,
Sugar,	,,	Oil,	per bot. (½-pt.)
Currants,	,,	Shears,	per pkt.
Raisins,	,,	,,	per pair.

Signed at   , this   day
    of   , 188   .

Witness—

———

The following is the form of an agreement for wages-men employed at shearing time :—

MEMORANDUM OF AGREEMENT MADE THIS   DAY
    OF   , 188 , BETWEEN   , OF
                                         AND THE UNDERSIGNED
WORKMEN.

Whereby the said workmen hereby agree to serve the said
                                 during the shearing season
    of 188 , at   station, and to obey the
following rules :—

1. That no money shall be paid to any man during shearing, nor any amount paid on his behalf, until the completion of shearing.
2. That the said workmen shall obey all reasonable commands given by the said   , or whom he may appoint to superintend them.
3. That no man shall leave the shed without permission, or until all is tidied up after every stoppage of shearing.
4. That the said men shall assist to wash the board, or do any other work they may be directed to.
5. That any man who is dismissed, or who leaves before

the completion of shearing, shall be paid off at the rate of   shillings per week.

6. In the event of wet weather, if too wet to shear, and dry enough for other work, the undersigned men shall go to any work that may be allotted to them.

Any of the undersigned workmen who perform the duties allotted to them well and faithfully, until the completion of shearing, will be paid by the said at the rate of   shillings per week.

Signed this   day of   , 188 ,
    at   station.

Witness—

CHAPTER XV.

THE WOOL-SHED.

THE wool-shed, like drafting yards, should occupy, as nearly as possible, a central position, for the convenience of bringing sheep to it from all parts of the run. Of course, where there are two sheds on one run, this rule does not hold good. In building a shed, a good one should be built at once, instead of the too common practice of squatters, of putting up temporary ones, the primary cost of which, added to the sums annually expended in repairs, will in a few years amount to more than the cost of a good substantial shed. Besides this,

the inconvenience which an inferior shed causes, and the
liability of the wool to be damaged by rain, dampness,
&c., is great; also, the loss of small quantities of wool
daily, which in a properly constituted shed would not
take place, amounts, at the end of shearing, to a considerable amount. We are of opinion that, however
small a shed may be required, it will be found cheaper
in the end, and more advantageous in the meantime, to
construct a good one.

There are many descriptions of sheds that find favour
with our squatters; some consisting of shearing boards
on either side, with the sheep in the middle of the shed;
others, with the board in the centre and the sheep on
each side. These are called single and double-boarded
sheds. There is also the T shed, which is largely used,
and is built in the shape of a T; hence its name. It is
built as though it were two buildings—the one forming
the cross of the T being used as a shearing board,
sheep and sweating pens. This part of the building,
and as much of the other as may be required for the
wool-classing, wool-rolling, &c., is built upon blocks
sunk 3 feet in the ground, and raised to about 4 feet
above the surface, except two rows of spars reaching the
full height of the building, and extending from end to
end, each row an equal distance from the centre. The
remaining portion of the other part of the building is
also built upon blocks, placed not more than 10 feet
apart, and sunk to the level of the ground; two rows of
spars extending from end to end, as in the other part of
the building. This portion is used as a pressing and

wool room, and, as such, requires extra strength in its foundation.

The shearing board extends the whole length, and is built upon the wool-room side of the shed. The pens built to contain the sheep before shearing are placed up the centre of the shed, either one to each or one to two shearers; behind these a narrow race is made, which is filled from the sweating, or night pens, which consist of the remaining portion of this part of the shed. Through two or more gates, the sheep are from there transferred to the shearing pens, as required. Upon the opposite side of the shearing board, " port-holes," or small door-ways, are made (one for each shearer), through which the sheep are turned when shorn, and conducted to the ground in their respective pens, constructed outside the shed, upon smooth boards or planks, in a gradient not exceeding 3 feet in 1 foot. The men shearing opposite the wool tables dispose of their sheep in a like manner— the pens being built on the other side of the shed, and carried underneath to the shoots.

That portion of the wool-room which is built at an elevation, and where the wool tables stand, can be laid with a second floor on the ground level, and boarded up all round, divided into two, so as to form two compart-ments—one for first and one for second pieces, which find their way through two trap-doors on the first or top floor. Two doors are left so that the pressers can have easy access to the wool from the lower floor. By this method the floor underneath and about the wool tables can easily be kept clear.

The wool-press must be built upon a solid foundation, and raised to the height of the ground floor, in a position that will give easy access to the wool bins.

The double-boarded shed, which has for its chief recommendation the splendid view obtained of all the men while at work, gives no man, however much inclined, a chance to "slum" his work without being observed by the manager of the shed. This shed we prefer to all others. It is built with about two-thirds of the whole length of the building on an elevation, as in the T shed—the remaining portion being built upon the ground level, and used as a wool-room. The shearing board should be made so as to allow plenty of room, from 15 to 20 feet, as overcrowding is dangerous. On each side of the board are built the sheep pens, which are filled from a race on each side, and extending the whole length of the board and along the back of the pens, and which is in its turn filled from the sweating pen, which occupies as large a portion of the shed as is admissible. The wool tables and trap-doors for first and second pieces are on the same principle as those we have already mentioned.

In constructing the inside pens of a shed, the divisions between the shearers' pens can be made either to draw back or to lift up (like a window), so that the whole can be more easily filled with sheep, when the shed is empty, than by forcing them into every separate pen through the gates. When the whole has been filled up, the divisions can be drawn down, and every man has his own pen.

The following is a good plan of yards to build in

conjunction with a shed, and one we have proved to work well :—

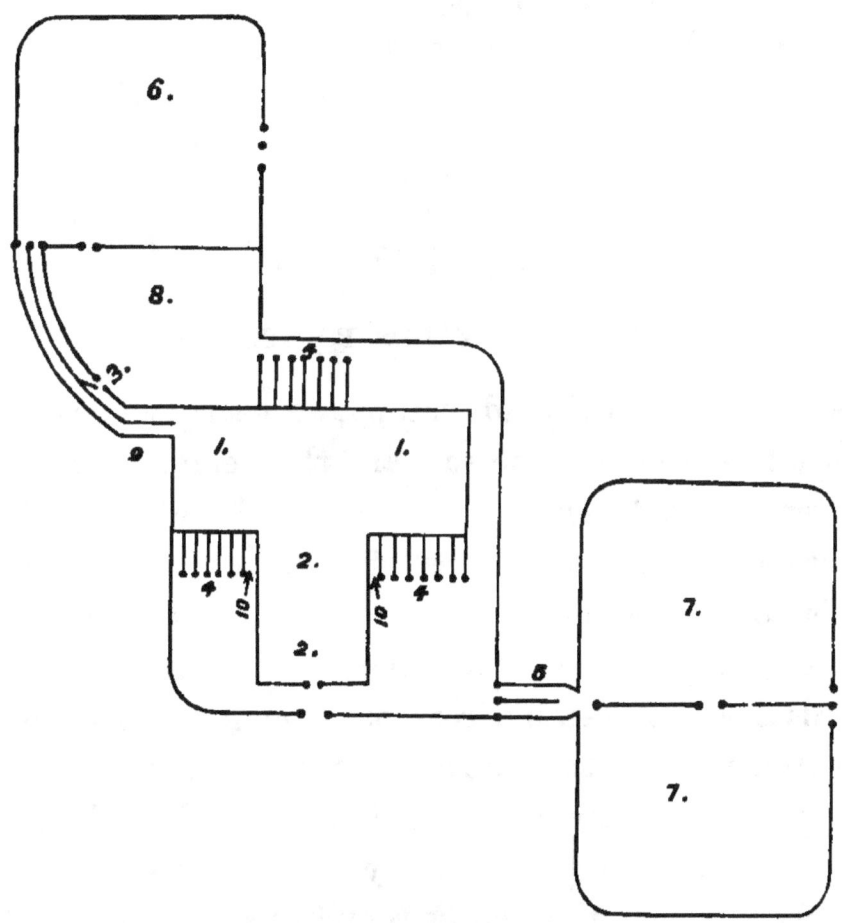

Plan of Yards, with Wool-shed.

1.1. Shearing board and sweating pens.

2.2. Rolling tables, press and wool room.

3. Drafting gate for taking off shorn or strange sheep.   This portion of the race is arranged so that a gate hung from the division of the race may close against a post in the centre, rendering the race of a suitable size for drafting, and, when not required, may swing back flush with the division, leaving no obstruction except the single post.   A second gate hung on the inside of the race will act as a drafting gate, and, when not in use, will, when closed, leave the race secure.

4.4.4. Shearers' pens.

5. Drafting and branding race.

6. Receiving yard for woolly sheep.
7.7. Receiving yard for shorn sheep.
8. Yard for strange and shorn sheep.
9. Entrance to shed for sheep.
10.10. Entrance to shed for men.

CHAPTER XVI.

DRAFTING YARDS FOR SHEEP.

DRAFTING yards should occupy, as nearly as possible, a central position on the run, and the nearer to water the better, as that necessary article will be found indispensable during the progress of drafting, and the nearer it is to be obtained the better. Round rails and posts have been proved to make the best fences for drafting yards, for innumerable reasons, amongst which may be mentioned strength, durability, and neatness. Although its erection is more expensive than that of the drop fence (which comes next in quality), stake fence, and many others, it will in the end be found cheaper, owing principally to the fact that the sheep suffer less, when coming into contact with it, from splinters (which are unavoidable in split rails), and from other objectionable features to be met with in the construction of various kinds of fences, the absence of which, in the round post and rail fence, renders it superior.

If pine timber is to be had handy, we would, of course, use all pine. The posts should be cut about 6 feet 6 in. long, and should measure not less than 8 in. in diameter at the small end, with the bark off. The rails ought to measure not less than 3½ in. in diameter at the small end, also with the bark off; the height of the fence may vary from 3 feet 10 in. to 4 feet 2 in., with four or five rails, according to taste and construction. A 3½ in. by 2 in. mortise will be found sufficiently strong, and should be bored from both sides of the post, four inches of which is quite enough to leave above the top rail. The expense of erecting such a fence as this will vary according to the locality, quality, and quantity of timber to be obtained, and the price of labour, from 4s. to 6s. per rod, without cartage. If the labour of a few hours be devoted to the removal of any knots that may exist on the *inside* of the three bottom rails, this fence will be found as nearly perfect as possible.

Where round timber is not available for either the last named, or drop fence, a stake fence (by some erroneously called stub) may be erected at about the same expense, viz., from 4s. to 6s. per rod, without cartage. Among the many ways in use for making this kind of fence we prefer that of digging a trench from 18 in. to 24 in. deep, in whatever direction it is intended to fence, and placing the timber in it, either upright or slanting outwards at the top, where a dog-proof fence is required. The trench should then be filled and thoroughly rammed. Extra strength may be procured in the dog-proof fence by placing a row of rails, resting in forked sticks, on the outside within a foot of the top.

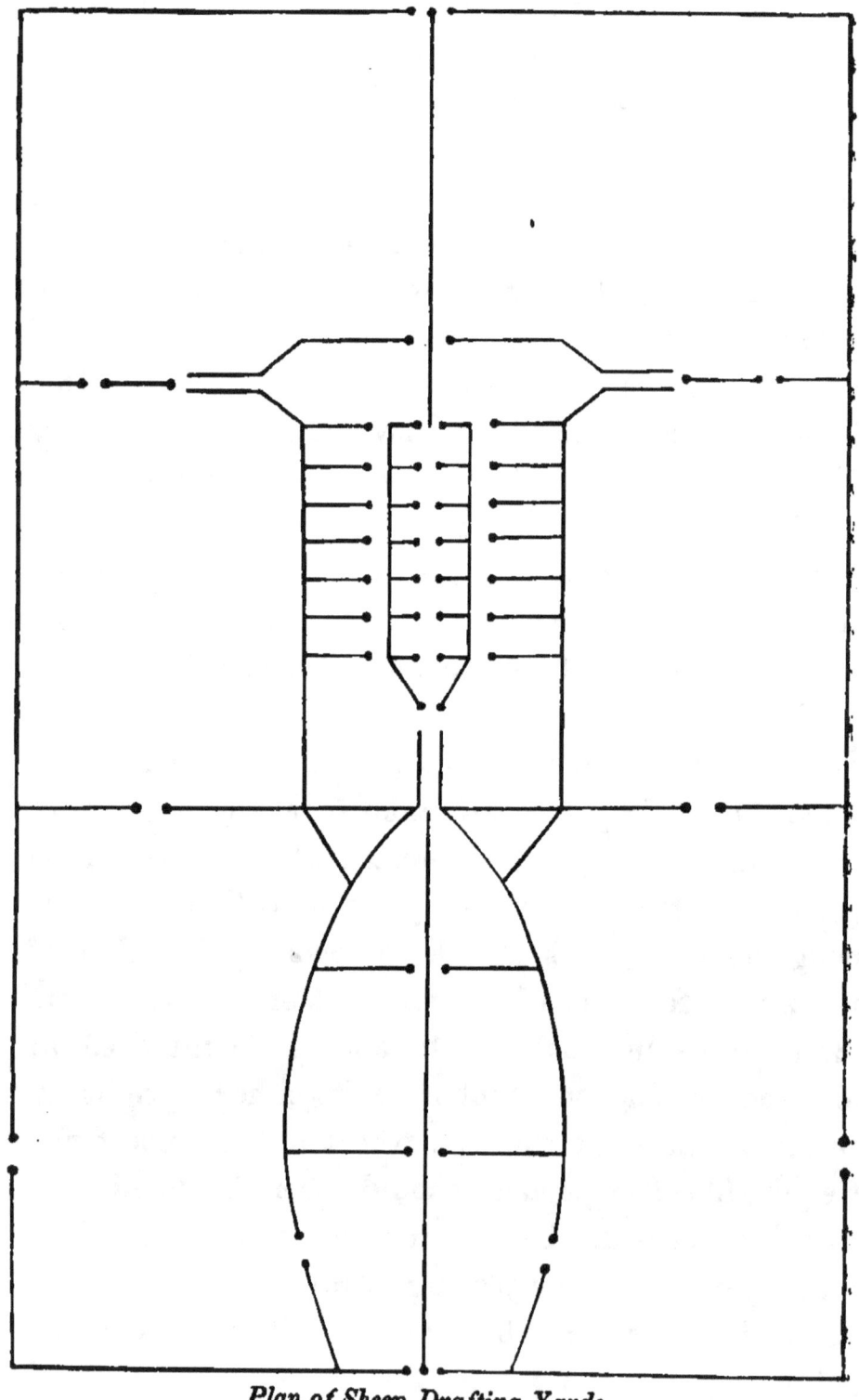

Plan of Sheep Drafting Yards.

These yards will be found very suitable for working large flocks of sheep. As will be seen, six different lots can be drafted by putting the sheep once through the yards. The small yards will be found very useful for lamb-marking or classing.

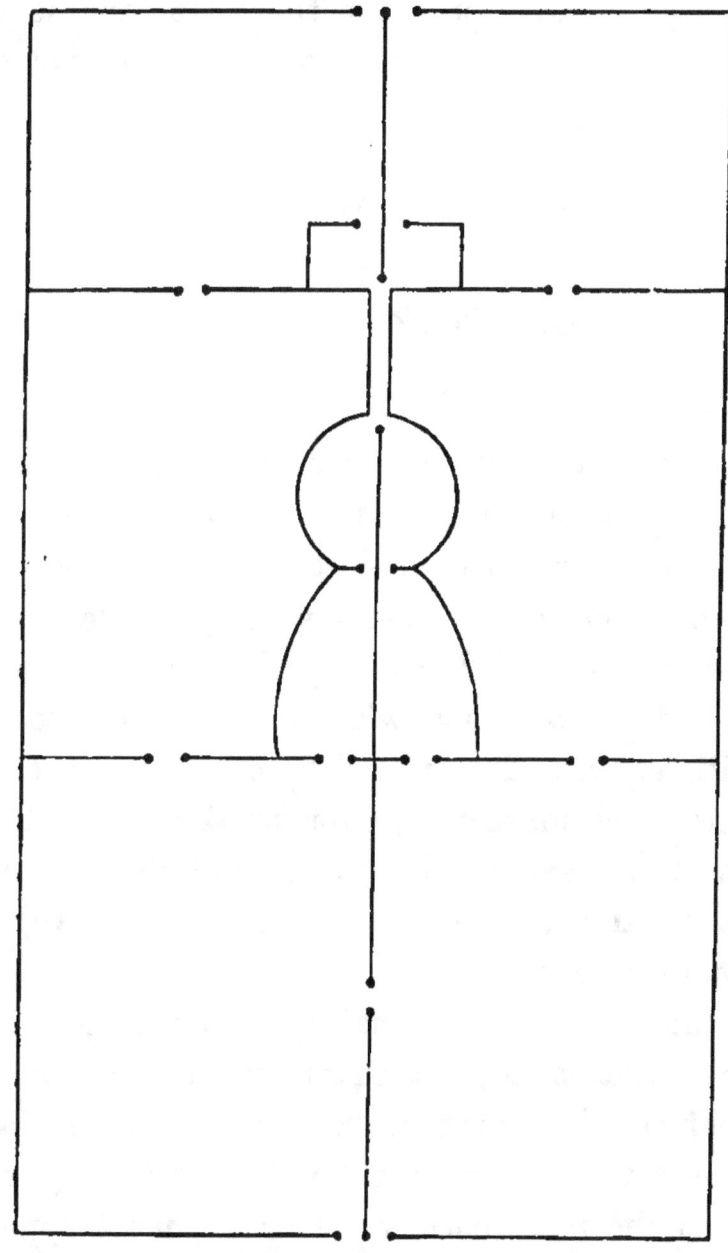

Sheep Drafting Yards for Small Station.

These yards are simple, and work well, and are very suitable for a small station.

We have refrained from entering into a long and detailed description of the yards and their working, because, to anyone to whom the plans are likely to be of any service, a description of their working is not necessary, and would occupy more space than we can devote.

CHAPTER XVII.

FENCING.

As, in the improvement of a run, fencing constitutes one of the most important items, we propose to enter somewhat largely into the subject; as, even among our squatting and managing community, great ignorance exists, and a want of knowledge is sometimes very apparent. We trust that what may appear to the better initiated an undue extension in our remarks will not be considered unnecessary; for, putting apart those who ought to understand the subject, and pretend to do so, there are many to whom a few useful hints will be of considerable value.

In the first place, many squatters and managers pay but little attention to the character of the country through which they propose to erect a fence; and, without regard either to this or the cost, they rush headlong at the wire fence as being the most popular, and, as some erroneously suppose, the most secure. By this injudicious practice many hundreds of pounds may, even on small holdings, be easily thrown away in the primary outlay, not to mention the cost of keeping these fences in repair. It would, therefore, be advisable that a thorough inspection of the country through which the fence is to pass be made by a competent person, with a view of ascertaining the description of fence which can be erected at the least cost and greatest suitability; and we would advise the construction of different kinds of fences on one line, where the description of country is unsuitable to one kind right through.

If our managers are to be censured for their lack of judgment and want of forethought in this respect, what must we say of the careless and negligent manner in which hundreds of miles of their division fences are laid out, both as regards want of straightness and proper direction? It has often come within our notice to see men, holding good positions on stations, and gifted with a large amount of faith in their own capabilities, commence to mark out a line of fencing, relying more on their personal knowledge of the run than on the correctness of the compass; in consequence of which they have soon found themselves proceeding in a totally wrong direction; and to remedy this evil, instead of making a fresh and proper start, they make an angle, regardless both of the appearance and the extra fencing this entails. Thus, many comparatively short lines of fences are seen with many bends and angles, much to the discredit of the manager. If the money wasted on the extra amount of fencing required could be gathered into a lump sum, we think we could live comfortably, if not luxuriously, on the interest of it.

In marking out a line of fencing we would recommend, in all cases where it is practicable, that the direction be always in one or other of the cardinal points, or parallel with the boundaries, and that all angles should be right angles, or as nearly so as possible. Where a theodolite is not obtainable, or not understood, a prismatic compass is the best and most accurate instrument to use. After having obtained the direction in which it is wished to run the line, by placing two pegs in the proper course—which must be done with great care and accuracy, as, should either of these deviate in the slightest degree from the proper line, it will be found that the deviation will be great at the end, according to the length of the line—nothing but a straight eye and care is required to bring about a satisfactory result. In marking the line, any timber that prevents a clear sight should be cut down; but where this is not desired nor convenient, the following plan will be found useful and correct for passing a tree that may be in the way, and is termed a set-off. Having arrived at the obstruction, two pegs are placed, say ten yards apart, one being close to the line; then place two more pegs at right angles to the last two, equal distances apart; we thus have a start for a fresh line running parallel with the first one, and clear of the tree. When the tree has been passed on this line, the original one is resumed by the same method. Straight pegs should always be used, and when not to be obtained on the line, a sufficient number should be split before starting, and distributed from a cart as required. They should not be placed too far apart—from 30 to 40 to the mile.

WIRE FENCING.

The posts to be employed in erecting a wire fence should be from 3½ x 7 in. to 4 x 8 in. in size, and from 4 feet 6 in. to 6 feet long. Where box, or any other timber of the eucalyptus species, is used, the posts are generally split and barked; but in many parts of Riverina, where the timber is only procurable from swamps, and is consequently small, this is not practicable, and it is therefore used in the round. When it is desired to bark the posts, that portion which is sunk in the ground is quite sufficient, as what remains exposed to the weather will, in the course of a few months, detach itself from the posts, and is thus removed without the expense of barking. The posts should be sunk to a depth of from 20 to 24 in. in the ground, and a straining post placed at from every 50 to 70 yards, and sunk 3 feet in the ground, with a stay from each side of the top of it to the bottom of the next "running" post. These stays are to be bored, and all the wires run through it before being fastened to the straining post. The further the posts are placed apart, the less the cost of erection. The distances vary from 9 to 24 feet. When the distance is above 9 feet, braces should be used by fastening a wire vertically round each of the wires excepting the top one, already strained, the number being regulated by the distance between the posts, the width of the braces being generally from 4 to 5 feet. We have proved a fence of 9 or 10 feet panels, without braces, to excel any other wire fence, and to be much more easily repaired, as a broken wire can never be properly re-strained when it is bound by braces at stated intervals. Where timber is scarce, it sometimes becomes impracticable to place the posts at these distances ; indeed, we have known instances where the timber for wire fencing had to be carted, by teams, distances from fifteen to twenty miles, and in many parts of Riverina nine and ten miles are common distances from the nearest timber. Of course, in this description of country, no other than the wire fence can be erected, and wide panels become almost a necessity ; and, where this is the case, braces must necessarily be used. The holes in the posts of a wire fence should be bored with a $\frac{1}{2}$-in., or not larger than a $\frac{5}{8}$-in. auger. The following gauge will make a neat and secure six-wire fence :—From ground to bottom wire, 6 inches ; from bottom to second wire, 5 inches ; from second to third wire, $5\frac{1}{2}$ inches ; from third to fourth wire, 7 inches ; from fourth to fifth wire, 8 inches ; and from fifth to sixth wire, 11 inches. This gives a fence 3 feet $6\frac{1}{2}$ inches in height, which will be found excellent for sheep, and quite as good for cattle as a wire fence can be made. This fence can be made still more sheep-proof, and no more costly, by leaving out the bottom wire, and having the second one at a distance of 11 inches from the ground, and a light embankment thrown up to nearly this height, thereby rendering the fence less liable to destruction by fire. This must, of course, be done during the winter, or wet weather, when the ground is soft, and will not answer in stony country.

Should a higher fence be required, the bottom wire may be raised to a height of 8 inches from the ground, and the other wires proportionately, and a plough furrow thrown up on each side of the fence. This also acts as a check in case of bush fires. For those who require a five-wire, and, consequently, a lower fence, the following gauge will be found suitable :—From ground to bottom wire, $6\frac{1}{2}$ inches; from bottom to second wire, $5\frac{1}{2}$ inches; from second to third wire, 6 inches; from third to fourth wire, 7 inches; and from fourth to top wire, 12 inches. This gives a fence 3 feet 1 in. in height, and sheep-proof. The gauge of a wire fence is a matter of taste, and admits of various opinions; but we have proved those we have given to be excellent in every respect.

ONE-RAIL AND FIVE-WIRE FENCE.

This fence possesses the advantage over the ordinary wire fence of being cattle-proof, or nearly so, and makes a very suitable boundary fence where timber is somewhat scarce. The rails may consist of any timber, either split or round, that is obtainable, the posts being a little stronger than the wire-fencing posts—say not less than 8 x 4 inches—and, if possible, of tough, durable timber. They may be either round or split, according to what is available. The barking is, as we have before remarked in reference to the wire fence, a matter of opinion, but, where desired, is, of course, subject to those remarks we have already passed. The following gauge will give satisfaction :—From ground to bottom wire, 5½ inches; from bottom to second wire, 6 inches; from second to third wire, 6½ inches; from third to fourth wire, 7 inches; from fourth to fifth wire, 8 inches; from fifth wire to top of rail, 14 inches, thus giving a total height of 3 feet 11 inches. This description of fence is one which is largely used by the squatting community, and has many advantages to recommend it; although it is not *entirely* cattle-proof, it is less likely to be broken by them than the ordinary wire fence. Our remarks relating to the throwing up of a light embankment, or plough furrow, are applicable to this as well as the former and other wire and wire-and-rail fences yet to be mentioned.

TWO-RAIL AND FOUR-WIRE FENCE.

This fence is both cattle and sheep proof, and, when properly constructed, forms as good and substantial a fence as any squatter or selector need desire. We will give, further on, the approximate cost of the various fences we are describing, by which means those about to erect fencing may ascertain the probable outlay required, bearing in mind our former remarks relating to the description of country. By this table it will be seen that the two-rail and four-wire fence is more expensive than any we have yet described, but is, of course, more substantial than the ordinary wire fence, and possesses, in common with many other descriptions of fence, the advantage of requiring less attendance when once erected. The posts should be sunk to a depth of not less than 2 feet in the ground, and, if of tough timber, should have a length of 4 inches above the top rail, which by the following gauge will be seen to be 4 feet 6 in. from the ground. Should the timber not be of tough quality, the distance may be increased to 6 inches. They should be bored at distances of 6, 5, 5, and 5 inches, starting from the ground; from the top wire to the top of bottom rail, 16 inches, and from the top of bottom rail to top of top rail, 17 inches. Care must be taken in the boring, and the exercise of a very little forethought will show that the posts must be bored before the rails are placed. The change of work which must take place in the erection of this fence increases the liability of irregular boring more than in the wire fence, where the boring is straight ahead work. Staples are sometimes used to fasten the wires on either the inside or outside of the posts, but we cannot recommend them as being either more secure or less costly than the boring. To render this description of fence, if possible, more cattle-proof, the bottom rail may be lowered a few inches and the fourth wire removed and placed midway between the top and second rails.

THREE AND FOUR-RAIL FENCES.

These fences, except for small paddocks and yards, are very seldom used on stations, principally because of the extra time occupied in the erection of them, and the heavier outlay required to complete any quantity. We will, therefore, content ourselves with but few remarks upon them, and pass on to fences more suitable to the requirements of the present time. No rail fence should have more than 9 feet 6 in. panels, and unless they are well and substantially constructed, it is better that they were not erected at all. The posts should be, if split, at least 8 x 4 in. in size, and sunk into the ground a depth of 2 feet. The rails, if split, should be from 9 x 3 in. to 7 x 3 in., and, if round, 4 inches in diameter at the small end. The mortises depend somewhat upon the manner employed in fitting in the rails, which is sometimes done by cutting the rails in a slanting direction, the cut being the length of the width of the post, and placing one upon the top of the other, which necessarily requires a wider mortise, to give the requisite strength, than the old and time-honoured plan of shouldering the rails and placing them side by side in the mortise, fitting tightly and completely filling every crevice. In the making this, or any other kind of fence, neatness may as well be studied, and this cannot be done where the mortises are chopped out with a mortising axe; therefore, the use of a two-inch auger is recommended. Although the former process may be somewhat quicker, it is not nearly so strong or neat. The distances between the mortises depend entirely upon the height of the fence required, and the taste of the person erecting it.

CHOCK AND LOG FENCE.

This fence we consider the most useful, secure, substantial, and economical of all fences for general station purposes, and we would strongly recommend the erection of it in preference to all others wherever it is practicable. We have used the word economical advisedly, for, although there are fences which can be erected at one-half the primary cost of the chock and log fence, we are quite certain that, at the expiration of a few years, the latter will be found to have filled all requirements, and to have sustained but little injury from natural sources; while the former, and wire fences, will have been a constant source of annoyance and expense, caused by the destruction liable from kangaroo, emu, and other animals. Nor is the chock and log fence any more likely to be destroyed by fire than any other fence, except, perhaps, those protected by an embankment such as we have described. Where suitable, this fence is less costly than the wire, wire-and-rail, or rail-and-post fences; and we consider it equal to any of them, with, perhaps, the exception of the post and four-rail fence, but the immense difference in cost renders it preferable even to that. The longer the panels of the chock and log fence, the cheaper it is; and the longer they are the worse it is, as the extra length of the logs causes them to "sag," and, in time, break in the middle. A very desirable length for the panels of a chock and log fence is 16 feet 6 in.; and no panels should be made longer than this. For a four-log fence of a good, substantial character, the bottom log should be from 8 to 10 inches in diameter at the small end; the second one, 6 inches; the third, 5 inches; and the top log, 4 inches in diameter at the small end. The bottom chock should be 10 inches; the second, 8 inches; the third, 7 inches; and the fourth 6 inches thick. This, after deducting the depths of the hollows which are cut for the logs to rest in, gives a fence of a little over 4 feet in height. By increasing the size of the logs and chocks, the number can be reduced, and the fence still be the same height. In country of a swampy description, or which is liable to become boggy during wet weather or winter time, a small piece of split timber should be placed under each end of the bottom chock, to prevent sinking. In the erection of a fence, split timber may be used; but, if possible, all round timber should be employed for the logs.

DROP FENCE.

This is a secure, but rather expensive fence, and is one we would not recommend for general purposes, although it makes very good yards, as will be seen by reference to that subject. It is better, neater, and lasts longer when made with round rails and split posts. It makes excellent " gate-marks," about four or five rods being placed upon either side of gate-ways in wire fences. It is made by placing two posts (half-round, or D-shaped, being the best, as they are enabled to fit close against the rails), not more than 4 inches apart, in each hole, both being well rammed, and sunk to a depth of not less than 2 feet in the ground. It is better to erect a number of posts before commencing to place the rails, which is done by cutting the ends of them so as to make them fit between the posts—a small block being laid between every two posts, underneath the bottom rail, to prevent it sinking too close to the ground. When the rails are placed, the posts must be securely bound together with a double ply of wire, either just above the top rail or sufficiently low to allow this rail to rest thereon. Should it be desired to bark the timber in this fence, the posts must be completely barked before being placed in the ground, as the wire binding prevents the bark falling off, as in other fences.

ROLL-OVER OR BRUSH FENCE.

The only difference existing, of any consequence, between the roll-over and brush fences is in the fact that the latter can be constructed where the former would be found impracticable, owing to the difference in the timber required. The roll-over requires straight, "umbrella-shaped" timber; by this we mean that kind of tree which, after rising straightly to a certain height, branches outwards, with somewhat straight limbs, in almost every direction, like the wires of a half-closed umbrella. After a tree of this description has been brought to the line, always by bullock-teams, a man with an axe half cuts through those of the limbs which protrude the most. When this is done, a chain is fastened to the undermost limb, and carried over or round the tree; the bullocks are then fastened to it, and the tree rolled on to the line—hence the name "roll-over." As the tree rolls, those limbs which have been nicked, or half cut through, break with the pressure upon them, and fall into place, thereby forming a close, compact, and secure fence. The second tree is treated in the same manner, and rolled on to one-half of the trunk of the first tree—thus making a double foundation, formed by one-half of the trunk of the first and the whole of the trunk of the second tree, the limbs of the

14

last tree always covering up the butt of the second preceding one. It will be found more advantageous if four
bullocks can be employed for the express purpose of
rolling over, as a long team proves unwieldy, and causes
much time to be wasted.

The brush fence requires no such distinctive character
of timber, and can be built of almost any kind, crooked
or straight, the limbs being cut off and the trunk rolled
in on the line, the limbs and leaves being inserted into
any open space which may exist.

STUB FENCE.

This is a description of fence somewhat similar to the
latter of the preceding two, but of a more substantial
character. It is formed by placing the butts of two
trees side by side, from 18 inches to 2 feet 6 inches apart,
resting at one end on the ground, and at the other upon
a chock placed crosswise at some distance from the
lower ends of the two preceding logs, thus making the
foundation or body of the fence; the limbs are then cut,
the straighter the better, and placed, either upright or in
a slanting direction, in the space formed between the two
logs.

ZIG-ZAG FENCE.

This fence resembles the chock and log in its formation,
but differs from it in having no chocks. We have seen
the option given to the contractor of erecting either of
the two at the same price; but, as far as we are concerned, this choice would never be allowed. The name

of the fence is quite sufficient to explain its shape, and it is so formed as to allow the end of one log to form the chock, or rest, for the next. The objection to this fence is, that if one top log should happen to become displaced, it will throw off the next, and so on, for, perhaps, the whole length of the line.

LOG FENCE.

This is a fence seldom used on stations except where clearing or grubbing is done, the logs only being dragged in and built closely together; which, when properly done, forms a secure and neat fence, and is a good method of disposing of the timber, which would otherwise have to be burned.

DOG-LEG FENCE.

The dog-leg fence is now almost out of use, and is very seldom seen. It forms neither a secure nor neat fence, and is so seldom used that we consider an explanation of its construction almost unnecessary. A somewhat similar description of fence may be used to raise the height of a low chock and log, which is done by placing two straight spars in opposite slanting directions at stated intervals, resting against the top log, and then placing additional logs in the forks formed by these spars. A two-rail fence may be formed by placing one log upon sawn blocks, or in forks, sunk into the ground, and about 2 feet from the surface, and completing the fence in the manner we have just described.

BRUSH DROP FENCE

Is suitable only to scrubby country, and is commenced, as in the drop fence, by placing posts, of any description, about 1 foot apart, with panels suitable to the length of the scrub to be used. Dry yarren makes the best fence. The scrub is dropped between the posts to whatever height may be required, the posts being then tightly bound with wire. This fence can be built with forks instead of posts, cut and sunk into the ground, and then filled up with scrub; this method does not make so good a fence, but saves the expense of the wire required to bind the posts. For a very temporary and inexpensive fence the following will be found serviceable:—Place posts at long distances apart, say 20 yards, or even further, if necessary; one hole is then bored at the top, and a strong wire strained tightly through it, and green bushy boughs hung thereon. This fence can be removed and the posts used again in the ordinary wire fence.

APPROXIMATE COST OF FENCES.

Upon the labour market, and the description and quality of timber convenient, depends solely the cost of fencing. When the difference in the price of labour which is now existing in various parts of the colonies is considered, it will be seen that it would be quite impossible to make any computation except by fixing a uniform rate of wages, which we propose to do at 30s. per week, with board, this being about as high wages as are paid by contractors to good men. It will, therefore, be necessary for those who may wish to benefit by our figures, to compare this rate with that in existence where they desire to fence. The following table shows the quantity of wire of the different sizes required per mile to fence :—

Cwt.	qrs.	lbs.				
6	3	17	of No. 4	wire reaches	one mile.	
5	3	6	„ 5	„	„	
4	3	14	„ 6	„	„	
4	0	13	„ 7	„	„	
3	1	23	„ 8	„	„	
2	3	15	„ 9	„	„	
2	1	12	„ 10	„	„	

To ascertain the quantity of wire required of any number, the figures we have given have only to be multiplied by the number of wires it is proposed to use, and to this sum must be added an allowance for splicing, and extra wire used in fastening to the straining posts. For example, a six-wire fence of No. 10 wire requires 14 cwt. 0 qrs. 16 lbs. per mile ; to this add 56 lbs. for the purposes we have named, making a total of 14 cwt. 2 qrs. 16 lbs. In wire fencing, the cost of posts, at the rate of wages we have fixed, is from 10s. to 12s. per 100 for pine posts such as we have described. For box or gum posts, the price is from 20s. to 40s. per 100, split; round box, or gum such as is found in swamps, from 12s. to 15s. per 100. Boring posts for six wires cost 10s. per 100; running and straining, 10s. per wire per mile; sinking holes and putting up posts, from 2½d. to 3d. per post, according to the depth and description of sinking. Thus we find that, taking the average of the above prices, a six-wire fence of 9 feet panels costs £17 per mile, without having allowed for the drawing in of the timber. According to this it can easily be seen what price per mile should be paid to any contractor, allowing him a fair amount for the use of his teams; and, supposing that the distance of cartage makes this allowance £4 per mile, we have £21 per mile as the contract price for the erection of the fence. To this add the cost, with carriage added, of 14 cwt. 2 qrs. 16 lbs. of No. 10 wire, and we have the total cost per mile.

The cost of the posts and rails for a one-rail and five-wire fence—of each of which there are, in round numbers, 600 per mile—is of the former about the same as the wire-fencing posts, viz., from 10s. to 12s. per 100, and of the latter, from 12s. to 14s. per 100. The total cost of erection, including running and straining the wires, is from 1s. 3d. to 1s. 6d. per rod, two pannels to the rod. Taking the average of these figures we have—Cost of posts, £3 6s.; cost of rails, £3 18s.; of erection, at 1s. 3d. per rod, about £20; making a cost of £27 4s. To this, again, add the price, with carriage added, of 17 cwt. 1 qr. 3 lbs. of No. 8 wire, with 3 qrs. added for splicing, &c.—say 18 cwt. in all—which gives the total cost of one mile. With these figures and examples, anyone should be able to arrive at approximate cost of any of these descriptions of fences.

The following table will give the approximate cost of the various other kinds of fences, and may be a guide to many whose lack of experience disqualifies them from estimating the probable cost :—

	£ £		£ £		£ £
Chock and log ..	30 to 40	Roll-over ..	20 to 22	Brush drop ..	20 to 30
Drop	35 to 40	Dog-leg ..	10 to 15	Log	20 to 25
Brush	18 to 20	Zig-zag ..	30 to 40	Stub	22 to 26

DESCRIPTION OF WIRE.

An immense amount of correspondence and newspaper controversy has been expended upon this subject; and, from what we have seen of it, we think the balance has been largely in favour of the Samson oval wire. At the time the Samson oval wire was introduced, the wire then in the market was of a very inferior quality, and the comparisons then instituted were greatly in favour of the Samson wire. We have, however, seen round fencing wire as good as any of the oval wire, either for strength or ductility. There is very little difference in the quantity used per mile, but what difference there is is in favour of the oval wire. The oval wire has other qualities to recommend it, one being that it will not break as easily, if run against by kangaroo, emu, etc., as the other wire, and when it does break it stands re-straining well. Although this wire is equal to all we have said of it, we do not think it will withstand the onslaught of *a mob of wild horses*, as some of its most enthusiastic advocates have affirmed. Anyone who knows what a mob of wild horses going at the rate they are often seen travelling at, especially when hunted—when their regard for a fence is much less than at any other time—will do to a wire fence, must know that nothing much less than the "great wall of China" would offer any resistance to their headlong career. We are not in favour of wire fences where the chock and log can be erected, but to those about to erect wire fences we would recommend the use of Samson wire—that is, as long as the manufacturers of this brand continue to supply our markets with as good a quality of wire as they are now doing; but should they follow the example of the manufacturers of many other brands, and, indeed, of many other articles—shears, to wit—and allow the quality of their manufactured article to deteriorate, when they have established a reputation and think they can foist an inferior article on the public, they will soon find the demand for their goods decreasing, and some new brand, in reality no way superior to what the old one was at the beginning, will spring up and take its place.

HURDLE FENCING.

On this subject the *Sydney Mail* has the following :—

" English hurdle fences have been in use for forty years or more, yet they show no signs of decay. The durability and desirableness of this kind of fence having been demonstrated, it remained for an American inventor to cheapen and perfect it, and to simplify its construction and to facilitate its transportation and erection.

" The fence is composed of flat iron bars and posts, alternate posts being provided with prongs or anchors. The horizontal rails are grooved longitudinally, to afford a seat for the rail. Wherever the rails lap, the mortise in the post is enlarged. When iron pickets are used, clips are employed to hold them. The picket passes through the holes in the clip, and the latter is fastened by a key or nail driven in between it and the ground side of the rail.

"To ensure great strength and steadiness, the posts are placed but 3 feet apart. The fence has a light appearance, but not too light, being readily seen by horses and cattle; besides, it is very stiff and strong. It has no barbs to injure stock; neither does it require straining posts or pillars. The rails and posts are sufficiently rigid to be self-sustaining. The fence can be graded or curved to suit any inclination. As to the matter of cost, it will compare favourably with the wooden fence; but when its durability is considered, it is found to be far cheaper."

FENCER'S AGREEMENT.

This agreement made this   day of   ,
188   , between   of the one part, and
                        of the other part. Whereby the said
                        hereby agrees to construct for the
said   about   miles of fencing,
according to the following specifications :—The fence to
be erected on a line commencing at   , and running
        miles. The posts to be placed at distances of not
more than   feet apart, and to be not less than 8 x 4
inches in substance; to be   feet in length, and to
be sunk   feet   inches in the ground. A
straining post to be placed at every   chains, and to
be not less than 8 inches in diameter at the small end,
and to be sunk 3 feet in the ground. A stay is to be

placed at each side of each straining post, extending from within 6 inches of the top of it to the ground, at each of the next running posts. Such stay to be bored and the wires run through before fastening to the straining post. The posts to be bored in the centre, and in accordance with a gauge supplied. All wires to be tightly and properly strained, and spliced with the knot splice.

The said   hereby agrees to complete this fence according to the abovementioned specifications, and to the entire satisfaction of the said or whom he may appoint to superintend the work. And the said   hereby agrees, in consideration of the fulfilment of this agreement by the said   , to pay the said at the rate of £   per mile.

(Signed)

Witness—

SPLICING WIRE FOR FENCING.

There are numerous methods of splicing wire for fencing purposes. We give below figures of what we consider four of the best kinds of splice. Fig. 1 represents the KNOT splice, now almost universally used; it is better

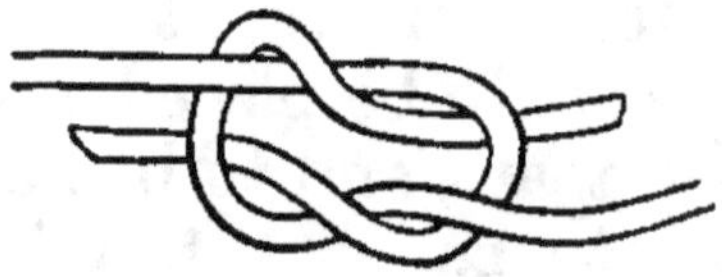

Fig. 1.

adapted to wire that is not pliable, and somewhat brittle, than any other.

Fig. 2.

Fig. 2 is the LOOP AND TWIST splice, often used, but not so good as the preceding splice.

Fig. 3.

Fig. 3.—The LOOP splice. This splice is objected to principally on account of its liability to break under sudden pressure, and the fact of one wire being liable to cut the other.

Fig. 4.

Fig. 4.—The TELEGRAPH splice. This is the oldest of all splices, and is one we would recommend only where good wire is used, as the short turns necessary to make it generally cause the wire to break, sometimes when one-half the splice is completed.

The knot splice is, in our opinion, the best and most secure; although not very long in use, it is gaining favour rapidly, especially in New South Wales. Splicing is a thing that should always be properly and carefully done. We have seen wires break as often as seven and eight times at the same splice during the process of straining; this, however, has been during cold, frosty weather, when it is always advisable to carry a tin of hot coals to warm the wire before commencing operations.

A tin such as is used by the tinsmith will answer the purpose admirably, and save a great deal of unnecessary labour and loss of time.

The above figures represent the wire before being strained.

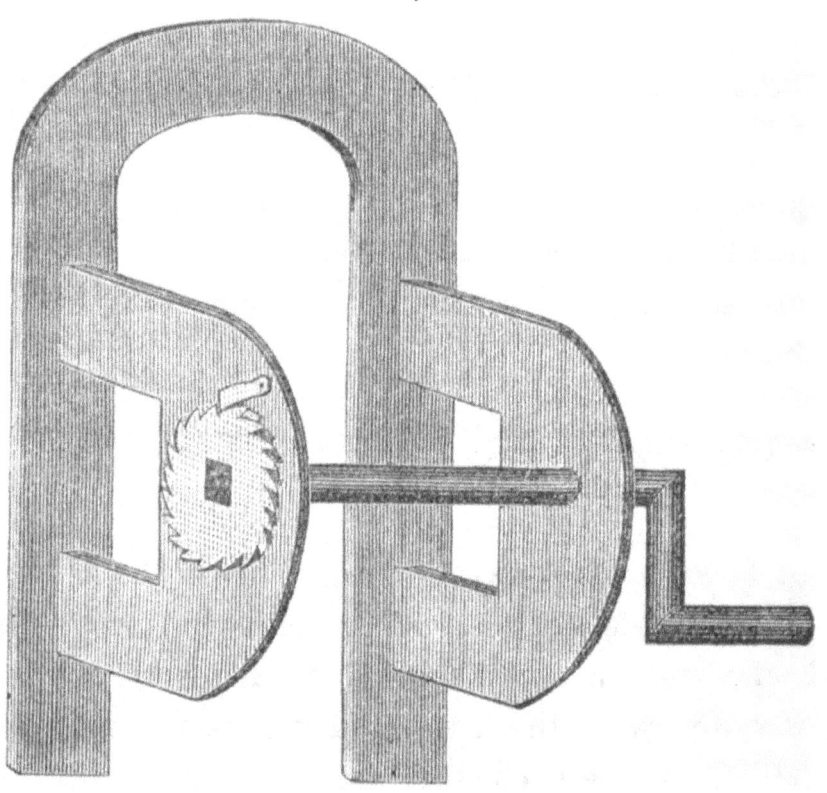

NOVEL WIRE STRAINER.

The above requires no explanation, as the drawing explains itself. This instrument will be found far more useful, and is more easily carried, than the old-fashioned Spanish windlass. It should be made of light iron, the roller being hollow if possible. Three short spikes, or legs, should be fixed behind, so as to give the instrument a grip of the post as soon as the wire is tightened.

CHAPTER XVIII.

DAM AND TANK MAKING, WITH RULES FOR THEIR MEASUREMENT.

To arrive at successful dam and tank making, every consideration must be paid to the choice of the locality, as well as to the description of ground in which it is proposed to make the excavation.

The construction of dams and sinking of tanks form by far the most important items of improvement where the country has not the advantages of a natural water supply. In the first place, trial shafts have to be sunk to the depth it is intended the tank shall be; and if the excavation proposed be a large one, more than one trial must be made, as it is sometimes impossible to judge, from one shaft, what the description of ground may be only a very short distance away. If the results of the first hole appear at all doubtful, the ground should be well tried before proceeding to excavate. Clay, or pipeclay, is, of course, the best description of soil for holding purposes, and is also excellent shifting stuff. Unfortunately, it is not often that this is found, as the sites of tanks depend, not upon where pipeclay, &c., is to be found, but upon good watersheds and accessible and suitable positions. These should be, as nearly as possible, in the centres of paddocks, so that the sheep may get to the water with the least possible travelling, and consequently less destruction of grass will take place.

The best slopes for the sides of tanks, for sheep, are not steeper than 2 feet in 1 foot, unless the tanks are fenced in and the sheep allowed to water at the roadway only, in which case slopes of 1 foot in 1 foot may be used, as the evaporation thereby is greatly lessened. Steeper slopes than these, however, we do not advise; as, where they are used, the sides frequently slide, and fall in; and although it certainly is true that the steeper the sides the less the evaporation, we are of opinion that 2 feet in 1 foot slopes are the best, as the quantity of water saved by the sheep's puddling— by tramping the sides of the tank, thereby rendering it more watertight—is greater than that gained by the decrease in evaporation where the sides are steep and the sheep fenced off.

Where, however, the slopes are made less than 2 feet in 1 foot, the sheep are able to enter too far into the tank, and thus cause great waste of water, in the quantities which they carry away in their fleeces; and to prevent this, hurdles may be arranged, at a sufficient distance from the water's edge, all round the tank, as to allow the sheep to drink, but yet to hinder them from going into the water above the knee.

The roadway should slope from the outside corners of the tank, and should have a slope of from 7 feet to 9 feet in 1 foot.

In sinking tanks, the object is, not only to get the excavation itself full of water, but to have as much "storage" room for water above-ground as the description of country will possibly admit of. Thus, if a tank be sunk between two hills, or elevations, the earth excavated may be placed so as to form a sort of dam, from the base of one hill, or elevation, to that of the other, and many feet of water saved which would otherwise have flowed to waste. Should such a position as this not be procurable, a site should be chosen as close to the steepest side of the hill as the sinking will admit of. If the description of soil renders it necessary, a puddle gutter should be made, commencing at the back of the tank, or that point furthest from the hill, and extending as far as the embankment; the latter should be built in the puddle gutter, and raised to any desired height, at the given point, and carried on that level, in both directions, until it strikes the hill, one end being made suitable for a by-wash.

Drains may be extended from inside the embankment, round either or both sides of the hill—their size being regulated by the size of the tank and the description of watershed.

The following description of tank will be found useful by those to whom the storage of water is a consideration. It is somewhat similar to the preceding one, but has some advantages over it, as will be seen. It is made upon flat country, near the base of a steep hill, and is of circular shape, earth being placed around it to any required height. A receiving tank is made upon the side of the hill, at a sufficient height above the level of the embankment of the circular tank as to allow of the free flow of water from one to the other, the fluid being conducted across by wooden fluming or troughing, placed upon shear-legs, a by-wash being formed upon the opposite side of the embankment of a similar size of troughing, so as to allow the water to flow out of the tank when full as quickly as it is conducted into it, and to fall clear of the embankment. Owing to the great depth of water which this description of tank may be made to contain, the evaporation will be less than is the case with ordinary tanks; it will, therefore, be found worthy of special consideration in hot climates.

In Victoria these circular tanks are largely made use of, but, instead of the troughing we have suggested, centrifugal pumps are used for filling the tanks, which are built in swamps. Culverts, or wooden drains running under the embankment, are sometimes used for allowing the excavation to fill, but we cannot advise the use of these, as their usefulness may be limited to one occasion, and there is no such limit to their liability to leak.

We have but to examine a copy of any of our weekly papers and journals to discover the lamentable ignorance existing among our population in regard to the rules for the measurement of tanks. Scarcely an issue passes without a score or two of replies to correspondents, indicating the contents of certain excavations, the measurements of which have been forwarded for calculation.

We therefore give the following clear and concise rules for ascertaining the quantity of earth excavated from any tank, the measurements of which are given; and we also select, at random, from the columns of one of our weeklies now at hand, several measurements that have been sent in for answers, which we will work out by the rules we give. These, we may mention, were published in that widely-spreading weekly, the *Town and Country Journal*, as long ago as the beginning of 1875 or end of 1874:—

RULES FOR THE MEASUREMENT OF TANKS.

Rule I.—To measure a circular tank, or cylinder.—Multiply the square of the diameter by ·7854 for the area of the top, and this multiplied by the depth will give the contents.

Rule II.—To measure a prismoid, or tank with sloping sides.—To the area of the top add the area of the bottom, and to this sum add four times the middle area. Multiply one-sixth of the sum total by the depth for the contents.

Rule III.—To measure the frustrum of a cone, or circular tank, larger at the top than bottom.—Add diameter of the top to diameter of the bottom; from the square of the sum subtract the product of the two diameters; multiply the remainder by one-third of the depth, and this by ·7854 for the contents.

Rule IV.—To measure an elliptical, or oval tank.—Find the area of the top by multiplying the larger diameter by the smaller, and the product by ·7854; or, multiply the same areas (or half the larger diameter and half the smaller) together, and the product by 3·1416 for the area. Find the area of the bottom in the same manner, and complete the calculation by No. 2 rule.

Rule V.—To measure the frustrum of a pyramid, or a tank with six, eight, ten, or any number of sides. —To the sum of the squares of the sides of the top and bottom add their product, and multiply their sum by the tabular number of the polygon, and again by one-third of the depth; or, add side of top to side of bottom; from the square of the sum subtract the product of two such sides, multiply the remainder by one-third of the height, and this last by the proper multiplier of the polygon for the contents.

EXAMPLE I.

To find the contents of a tank 167 feet by 82 feet on the top, 95 feet by 17 feet on the bottom, 9 feet 9 inches deep.

To find the top area, multiply—167 × 82 = 13694
,,   bottom ,,   ,,   95 × 17 = 1615

To find the middle area, add dimensions of top and bottom together, and divide by 2, thus—

$$167 \times 82$$
$$95 \times 17$$
$$\overline{2)262 \times 99}$$
$$\overline{131 \times 49\cdot6}$$

Multiply the product of the length by that of the depth, i.e.,

$$131 \times 49\cdot6 = 6484\cdot6$$

which is the mean area.

This multiply by 4—6484·6 × 4 = 25938
Add area of the top   13694
,,   ,,   bottom   1615
                                   ———————
                                    41247

This product divide by 6, and 41247 ÷ 6 = 6874·6
Multiply this quotient by the depth—6874·6 × 9·9 = 67026·9
This product divided by 27 will give—

The area in cubic yards—67026 ÷ 27 = 2482·12

The contents being 2482 cubic yards and 12 feet.

The contents of a tank 138 feet 6 inches long on the top, 126 feet broad on the top, 102 feet 6 inches by 90 feet on the bottom, and 12 feet deep; with a roadway 138 feet broad on the top at the end next the tank, and

90 feet broad on the bottom, carried out with a 7 feet in 1 foot slope to a width of 25 feet, are $7352\frac{18}{17}$ cubic yards. We append a diagram, with working of the calculation.

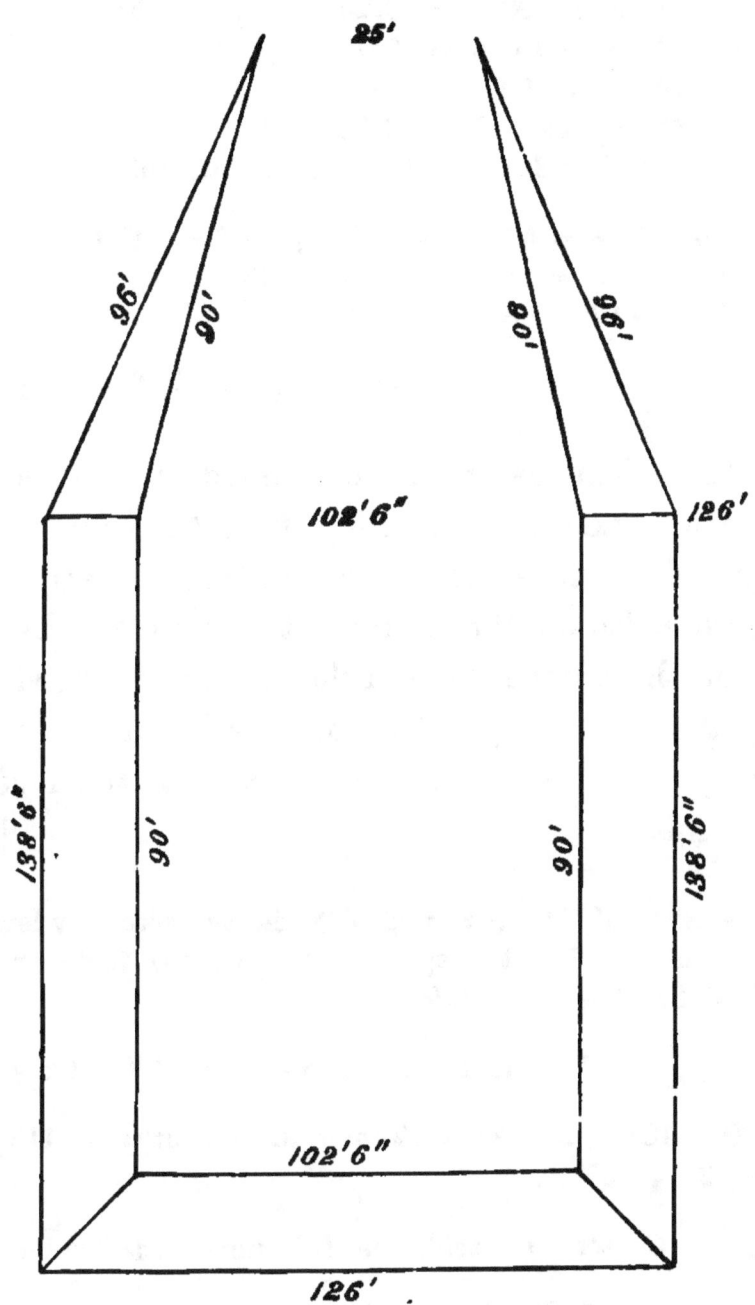

We proceed, first, with the calculation of the tank, by the same method as in the preceding example.   Thus :—

$$136' \ 6'' \times 126' \ \ = \ 17451 \text{ area of the top}$$
$$102' \ 6'' \times \ 90' \ \ = \ 9225 \quad \text{,,} \qquad \text{bottom}$$
$$138' \ 6'' + 102' \ 6'' = \ 241'$$
$$126' \ \ + \ 90' \ \ = \ 216'$$
$$241 \times 216 \div 2 \ = \ 120' \ 6'' \times 108$$
$$120' \ 6'' \times 108 \ \ = \ 13014 \text{ area of the middle}$$

$$13014 \times \ 4 = 52056 + 17451 + 9225 = 78732$$
$$78732 \div \ 6 = 13122 \times 12 = 157464$$
$$157464 \div 27 = 5832.$$

Contents of tank, 5832 cubic yards.

ROADWAY.—This has to be calculated in sections, the slopes being made up separately from the body of the roadway.   The rule is—for the body of the roadway, find the mean breadth on the bottom; this multiply by the length, and the product by half the depth ; for the slopes multiply the mean length by half the breadth, and the product by one-third of the depth, taking them separately.   Thus :—

$102' \ 6'' + 25 = 127' \ 6'' \div 2 = 63' \ 9'' \times 84$ (the necessary length, when the tank is 12 feet deep, with 7 feet in 1 foot slope) $= 5355 \times 6' = 32130 \div 27 = 1190.$

Contents of body of roadway, 1190 cubic yards.

$96 + 90 = 186 \div 2 = 93 \times 12$ (half the breadth) $= 1116 \times 4 = 4464 \div 27 = 165' \ 9''.$

Contents of each slope, 165 cubic yards 9 cubic feet.

Or, for both slopes, $330\frac{11}{17}$ cubic yards.

The whole contents, tank and roadway, being—

Body of tank	5332
,, roadway	1190
Slopes of do	330¼⁴
	——
Total	7352¼⁴

For a circular tank, or cylinder, apply Rule I., and supposing, for example, that we wish to find the contents of one 20 feet in diameter and 10 feet deep, we proceed as follows :—

To find the square multiply the diameter by itself, $20 \times 20 = 400$, which is the square of 20. This multiplied by ·7854 gives us 314·1600, which must be multiplied by the depth, $314 \cdot 1600 \times 10 = 3140$, and we may drop the decimal.

To find the contents in cubic yards divide by 27— $3140 \div 27 = 116\frac{8}{27}$, the answer.

To ascertain the contents of a circular tank, larger at the top than at the bottom. A tank, 15 feet in diameter at the top, 5 feet in diameter at the bottom, and 12 feet deep, contains $44\frac{5}{27}$ cubic yards, or, more accurately speaking, 44·2151. Thus :—

$15 + 5 = 20 \times 20 = 400 - 20 = 380 \times 4 = 1520 \times \ \cdot7854 = 1193 \cdot 8080 \div 27 = 44 \cdot 2151.$

An elliptical tank 30 feet by 10 feet on the bottom, 120 feet by 40 feet on the top, and 15 feet deep, contains 1112·6500 cubic yards, as the following working will show :—

$$30 \times 10 = 300 \times \cdot 7854 = 235 \cdot 6200 \text{ area of the bottom}$$
$$120 \times 40 = 4800 \times \cdot 7854 = 3769 \cdot 9200 \qquad ,, \qquad \text{top}$$
$$235 \cdot 6200 + 3769 \cdot 9200 \div 2 = 2002 \cdot 7700 \qquad ,, \qquad \text{middle}$$
$$2002 \cdot 7700 \times 4 = 8011 \cdot 0800$$
$$8011 \cdot 0800 + 235 \cdot 6200 + 3769 \cdot 9200 = 12016 \cdot 6200$$
$$12016 \cdot 6200 \div 6 = 2002 \cdot 7700 \times 15 = 30041 \cdot 5500$$
$$30041 \cdot 5500 \div 27 = 1112 \cdot 6500.$$

Contents, 1112·6500 cubic **yards.**

Or, multiply the semi-axis (or half the larger diameter and half the smaller) together, and the product by 3·1416, for the area, and complete the calculation by No. 2 rule, thus :—

$$30 \times 10 \div 2 = 15 \times 5 = 75 \times 3 \cdot 1416 = 235 \cdot 6200$$
$$120 \times 40 \div 2 = 60 \times 20 = 1200 \times 3 \cdot 1416 = 3769 \cdot 9200$$
$$235 \cdot 6200 + 3769 \cdot 9200 = 4005 \cdot 5400 \div 2 = 2002 \cdot 7700$$
$$2002 \cdot 7700 \times 4 = 8011 \cdot 0800$$
$$8011 \cdot 0800 + 235 \cdot 6200 + 3769 \cdot 9200 = 12016 \cdot 6200$$
$$12016 \cdot 6200 \div 6 = 2002 \cdot 7700 \times 15 = 30041 \cdot 5500$$
$$30041 \cdot 5500 \div 27 = 1112 \cdot 6500$$

After sinking a tank, it is always advisable to run drains into it; indeed, any saving of expense in this matter is seldom wise, as it frequently happens that the locality of a tank is visited by some passing shower, or thunderstorm, sufficient, if there be plenty of drains, to fill the excavation, and possibly of little or no good in the absence of drains. Many people apply the rule of 1000 yards of excavation to every mile of drain. For the calculation of the contents of drains the following scale will be found of use :—

DRAINS.

A drain 5 feet wide on the top, 3 feet wide on the
 bottom, and 2 feet deep, contains $17\frac{3}{37}$ cubic yards
 to the chain.

A drain 4 feet wide on the top, 2 feet wide on the
 bottom, and 1 foot 6 inches deep, contains 11 cubic
 yards to the chain.

A drain 3 feet wide on the top, 1 foot 6 inches wide on
 the bottom, and 1 foot deep, contains $5\frac{1}{2}\frac{3}{7}$ cubic
 yards to the chain.

A drain 2 feet wide on the top, 1 foot wide on the
 bottom, and 1 foot deep, contains $3\frac{1}{2}\frac{8}{7}$ cubic yards
 to the chain.

A drain 6 feet wide on the top, 4 feet wide on the
 bottom, and 2 feet deep, contains $24\frac{1}{2}\frac{2}{7}$ cubic yards
 to the chain.

Embankments for damming up water require to be
perfectly watertight, and to possess great strength. To
procure the former requisite, puddle gutters must be
sunk underneath the embankment and filled with clay,
and carried to the top, in the centre of the structure.

The great difficulty lies in preventing the water from
finding its way between the bottom of the puddle
gutter and the foundation to which it is sunk, and even
through the substances of which the foundation consists.
To prevent this, sometimes the gutter must be sunk to a
great depth below the surface, and in all cases until it
reaches an impermeable stratum.

Before commencing to construct a dam, a knowledge of
the geology of the locality fixed upon should be acquired,
and can only be obtained, as in choosing a site for a tank,
by sinking trial shafts.

The slopes of the embankment should be covered so as
to protect them from the action of the water; and
although, to effect this, various methods have been
employed, and have found their origins in circumstances
peculiar to the several localities in which they have been
tried, in most cases a coating of soil grown over with
grass will be found sufficient protection from the never-
ceasing action of the water, as the roots of the grass bind
the embankment.

In the case of a flood, nothing in the construction is to
be depended upon but the height and strength of the
embankment. In the event of the former being insufficient, and the flood water rising to a greater height than
the embankment (although the possibility of this should
never be allowed) and running over any portion of it,
bags should be filled with sand or earth, and immediately placed in such a position as to stop the running
of the water. This is a critical moment; and, if the
opportunity be not at once seized, the certain, and
perhaps total, destruction of the dam will follow. If the
water has too much play upon the end of the embankment, by the current being too strong, bags of sand will
be found invaluable to prevent the washing away of the
dam.

The possibility of this danger, however, depends
greatly upon the position of the embankment and the
nature of the by-wash.

Regarding the former, too much care cannot be taken
in the selection of a site. This should be chosen at that
point in the creek which will admit of the greatest

quantity of water being stored at the least possible cost, and where the banks of the creek are on a higher level than the country immediately surrounding them, so as to allow of a free get-away for the flood water; in fact, where the banks of the creek form a natural by-wash.

Where two such banks cannot be found, one will be sufficient, provided it be good enough to prevent the water from running round to the back of the dam, which would cut away the embankment. It is not often that all the natural advantages desired for the construction of a dam can be found in one spot, and that spot the one most suitable in regard to the relative positions of itself and other dams, tanks, and paddocks, all of which have to enter into the consideration of the question. But, if possible, a dam should be built at a narrow part of the creek, widening immediately above it, and possessing banks such as we have described.

An embankment should never be erected at a steeper slope than 2 feet in 1 foot, and if built at a greater slope than this it will necessarily be much stronger; but if it be made any steeper, the continual action of the water will cause the earth to slip. It should have a uniform width of not less than 9 feet on the top; and the height of it will determine, with the slopes stated, the necessary breadth at the bottom.

It should be made with a height of not less than 3 feet, at the highest side of the creek, at the end of the embankment, and carried on that level across; and the larger the embankment, the higher it should be above the level of the highest bank, so as to allow for the

sinking which will always take place when the dam first
fills and the soil becomes wet. This sinking, or shrinkage,
we have known to reach one-twelfth, and sometimes
even one-tenth, of the total height of the embankment.

In making the excavation, or tank, to procure earth
for the construction of the embankment, we would not
recommend its formation in the bed of the creek, as,
if there, it will invariably be filled with the sand, &c.,
washed down by the flood waters.

We have heard it argued that, if this were the case,
the whole of the dam would ultimately be filled up, and
that this would happen whether or not the natural state
of the bed of the creek was altered; but our experience
has taught us that the process of filling up extends only
to the excavation—that is, with any noticeable degree of
rapidity; and although in the course of many years to
come, during the lifetime of our great-grandchildren, our
dams may become filled up and useless, we consider the
contingency too remote to be worthy of consideration at
present.

Nor are we in favour of the too common practice of
breaking into and removing the natural banks of the
creek, because by this method a good coating of clay is
very often removed, which, if left as placed by nature,
would form an impenetrable barrier to any weight of
water. By the removal of this stratum, the water is
given access to any sand veins which may exist, and
which the puddle gutter may not cut, thereby causing a
leak; and we feel certain that in the majority of cases of
leaky tanks, the cause, if traced, would be found to be

the cutting of the natural banks, and thereby the defeat of the object of the puddle gutter which has cost so much to make.

The excavation, therefore, should be made at the end of the embankment, and at a safe distance from the edge of the creek, so that when the dam is full the water will run into the bank, and, after filling it, flow down the natural by-wash, and so return to its natural channel, the creek.

There are several methods of excavating, both for tanks and for the construction of dams, and all of them are considerably practised in various parts of the colonies. Some contractors use the plough and scoop, worked by bullocks; others plough the soil, using either bullocks or horses, and remove it by means of drays, drawn, of course, by the last-named beast of burden; and some more still retain the old method of pick and shovel work, disdaining the use of the plough, and excavating by means of horse drays.

For the construction of a dam we prefer horses and drays to bullocks and scoops (except where there are no sand veins, and the breaking into the natural banks of the creek becomes less objectionable than otherwise); but still we cannot deny the great advantage derived from the constant tramping of the bullocks in binding the soil of the embankment, and rendering it almost as solid as adamant.

In excavating a tank, however, where the soil is suitable, we give the preference to bullocks, with plough and scoop, for two good reasons—firstly, they make a better tank, more watertight and secure, by the tramping and puddling we have already referred to; and, secondly, the cost per yard is not so much when bullocks are in use as it is with horses and drays.

Of course, if the nature of the ground is too rocky to admit of ploughing to the full depth required for the excavation, the use of bullocks will be found impracticable: *ergo*, the result of the trials made upon the site before the commencement of operations may determine the nature of the working plant to be employed.

But even these trials are oftentimes most deceitful, and it has sometimes happened, after three or four most successful shafts have been sunk upon the proposed site of a tank, that, between the holes, considerable quantities of solid rock have lain; and, indeed, we have seen a trial hole sunk to a depth of 12 feet, which, if moved 18 inches in one direction, would have reached a stratum of solid rock at a depth of 7 feet from the surface.

Using the same rates of wages as we have done in giving the approximate costs of the different fences, we find that tank and dam making can be done at from 10d. to as high as 1s. 4d. per cubic yard by horse teams, either ploughing or using the pick, and depending both upon the character of the soil to be excavated and on the quantity of work to be contracted for; and at 8d. to 1s. per yard by bullock teams, using plough and scoop.

OVERSHOT DAMS.—This description of dam is seldom

seen upon stations; it requires no by-wash, and is, we think, deserving of a greater share of public favour than it now enjoys.

To erect an overshot dam, piles must be driven in, at equal distances, in a double row, across the creek—the closer the piles, the stronger the dam; the two rows being from 8 to 12 feet apart. Horizontals should be nailed or otherwise secured to the outside of the piles. Stays are placed from the top of each of the piles to the bed and banks of the creek, sloping from 2½ to 3 feet in 1 foot, upon either side of the dam.

If the creek is of a sandy nature, a puddle gutter must be sunk—between the two rows of piles, and must be down to a solid and watertight foundation—the full length of the structure, and the excavation refilled up to the top of the piles with good holding clay. The spaces existing between the stays and the piles must then be filled with the best clay procurable, and the whole boarded over with planks, securely watertight. A platform has then to be erected at the back of the dam, so as to carry the surplus water away from the immediate vicinity. The cost of such a dam necessarily depends upon its size, and the convenience of timber.

DAM OR TANK MAKER'S AGREEMENT.

This agreement made this day of , 188 , between of the one part, and of the other part, both of . Whereby the said does hereby agree to complete a dam and tank for the said and to his entire satisfaction, or to the satisfaction of whom he may appoint, and according to the following specification :—

The earth to be excavated from a place as marked, and placed upon a puddle gutter as mark   ,   feet wide, and sunk to a solid and watertight foundation, and filled with the best clay procurable and well rammed.

The embankment to be   feet long, with a uniform width of   feet on top, and with slopes of   feet in   foot. The excavation to be made with slopes of   feet in   foot, with a roadway at one end (or otherwise), and sunk to a depth of not less than   feet.

And the said   does hereby agree, in consideration of the said   completing the above work according to these specifications, and to the entire satisfaction of the said   or whom he may appoint, to pay the said   at the rate of   per cubic yard of earth excavated, and at the rate of   per cubic yard of puddle gutter.

(Signed)

Witness—

CHAPTER XIX.

WELLS.

IN some parts of Australia, wells constitute almost the only means of obtaining a sufficient supply of water to meet the demands of the stock ; owing to the level nature of the country, and the absence of watersheds, the enormous expense of conducting water, on level and sandy country, into tanks, and the extra quantity of rain necessary to fill these, render wells more suitable. Besides these, the immense advantages of a permanent supply of water, in case of drought, is quite beyond calculation. To attain this, efforts are constantly being made to obtain well supplies; and although in many instances these efforts have been unavailing, either by being unsuccessful altogether in obtaining water, or, when obtained, being too brackish for use, and sometimes too salt, yet there are many wells, in different parts of the colonies, whose owners need never fear the effects of a drought. In some cases, as we have already stated, water has been found too brackish and salt for use; but, worse than this, it has come under our personal observation to have seen water, found at a depth of 80 feet, and, to all appearances and taste, everything that sheep could desire, but which was evidently permeated with some mineral poison, and which caused sheep partaking of it to die a lingering death. For this reason we consider it advisable to have the water of *all* wells analyzed, or, if this should not be convenient, to allow but a few sheep to water until the purity of the water has been demonstrated. We say the water of *all* wells advisedly; as one poisonous one, that came under our notice, was surrounded, at distances varying from five to fifteen miles, by wells yielding, in all instances, a constant supply of good stock water, and in many instances quite fresh. By this it will be seen that the poisonous water is quite local, and good water may be obtained much closer than the distances we have mentioned.

It may be interesting to some, and instructive to others, to know that stock will sometimes at first refuse to drink water that they will eventually thrive on. As an instance of this, we know of a well on the Mirool, New South Wales, sunk to a depth of 296 feet, at which depth there was a good supply of water obtained, and which, having been analyzed, was proved to contain no poisonous matter, and tasted as fresh to the human palate as could be expected from water obtained at such a depth. In spite of this, however, stock, when put to it, refused to drink, and continued to do so for nearly two weeks, by which time they presented the appearance more of walking skeletons than of sheep and cattle. At the end of this time, one would almost suppose the water in the troughs to have been changed, from the sudden and ravenous manner in which it was drunk by the stock that looked the day before as if they must perish for the want of it. They afterwards came to water regularly, and fattened in almost as short a time as they had previously fallen away.

We may mention, among the causes of water out of new wells being refused by stock, the green timber used in timbering the shaft. Some people, who are not aware of this, erroneously imagine that the taste imparted to the water by the timber (especially pine timber) is mineral, and deleterious to the character of the water.

Among the peculiarities that have come under our notice, we may mention the following:—A well was sunk on a well-known station, not 100 miles from the Murrumbidgee River, to a depth of about 90 feet, when a small spring of very brackish water, unfit for stock, even had there been a supply, was struck; this was puddled back, and the sinking proceeded with to a further depth of from 50 to 60 feet, when a good supply of comparatively fresh and excellent stock water was obtained. Many persons would have discontinued operations upon meeting with the first spring; but in this case the indefatigable owners persevered, and were rewarded by the saving of thousands of pounds during the drought.

But little attention is, as a rule, paid to the aspect of the country when sinking a well, and too much, generally, to its position on the run. Both these are considerations of much importance, but the former deserves the greatest study; it is doubtless of more importance that the well should be sunk where the chances of obtaining water are greater; and when this is done, and success assured, it is an easy matter to divide the run so as to make the well available for as many paddocks as it is desired.

ARTESIAN WELLS are perpendicular borings into the ground, through which water rises from various depths, according to circumstances, above the surface of the soil. The possibility of obtaining water in this way in a particular district depends on its geological structure. All rocks contain more or less water. Arenaceous rocks receive water mechanically, and, according to their compactness and purity, part with a larger or smaller proportion of it. A cubic yard of pure sea-sand can contain, in addition to the quantity of dry sand which

16

occupies that space, about one-third of its bulk of water. It would part with nearly the whole of this into a well sunk in it, and regularly pumped from. Chalk and other rocks, composed of fine particles closely compacted together, contain as large a proportion of water; but, from the power of capillary attraction, little or none of this water would be drained into a well sunk in such rock. From the existence, however, of numerous crevices in chalk, through which the water freely flows, and from the general presence of a larger quantity of water than the porous rock is able to retain, wells sunk in chalk often yield water. There is yet a third class of rocks which are perfectly impervious to water, such as clays, which are absolutely retentive, neither allowing water to be obtained from them nor to pass through them. When such rocks occur in basins in alternating layers, and in such order that pervious beds are inserted between impervious ones, it is evident that if a perforation is made through the retentive barrier-bed in the lower portion of the basin, the water contained in the water-logged strata will rise through the bore to a height depending upon the pressure of water which has accumulated in the confined sloping space between the two impervious beds. The most famous artesian well, perhaps, is that of Grenelle, in the outskirts of Paris, where the water is brought from the gault at a depth of 1,798 feet. It yields $516\frac{1}{2}$ gallons in a minute, which is raised with such a force as to be propelled 32 feet above the surface. The pressure required to effect this has been calculated to exceed 50 atmospheres at the bottom of the bore. The water has a constant temperature of 81.7° F.

It is believed that the Chinese have been long acquainted with artesian wells. They have been in use for centuries in Austria, especially in the neighbourhood of Vienna, where they are very abundant. No knowledge existed as to their source, and, consequently, the boring for them was engaged in and conducted in a rude and empirical manner. An excavation was made till a bed of clay was reached; on this a perforated millstone was laid, and through the hole the clay was bored until water rose. As soon as geology took the position of a science, and the theory of artesian wells was propounded, the engineer was able, after the geological survey of a district, to discover whether a supply of water could there be obtained in this way. Already, districts formerly dry and arid have received a plentiful supply of water by means of such wells, and many more applications have yet to be made; it seems likely that ere long Africa's deserts may thus be converted into fertile plains. In an official report of the Algerian Government for 1856-1857, it is stated that artesian borings had been executed in the Sahara of the province of Constantine, with remarkable success. The first attempt, after a few weeks' labour, produced a constant stream, forming a perfect river, and yielding 4,010 quarts of water per minute, at a temperature of 78° F. There are now upwards of 75 such borings in the Sahara, yielding an aggregate of 600,000 gallons per hour. The result is proving beneficial, not only to the country materially, but also to the character and habits of its nomadic Arab inhabitants. Several tribes have already settled down around these wells, and, forming thus the centres of settlements, have constructed villages, planted date palms, and entirely renounced their previous wandering existence. There are several very deep borings in the United States.

To ERECT A WHIM.—In erecting a whim, the principal objects to attain are strength and durability, and it should therefore be constructed of some strong timber such as red-gum, ironbark, &c. The horse team should be, if possible, of one piece, but if not, can be spliced at the centre piece. The "collar trace" should be built sufficiently above the surface to allow the water, when raised, to flow freely into the tank built for the purpose of receiving it from the bucket, and to allow of the horse-walk being raised a few feet above the surface. The spare beam should be securely bolted and stayed at each end, the stays extending to within a few inches of the drum, which should be strongly made, about 9 feet in diameter, and of hard wood; the larger the drum, of course, the quicker the bucket is raised to the surface, although much more severe on the horse. Self-acting buckets are generally used, and are preferable to canvas ones, as the latter require a man to empty them. A whim when completed should have a good coat of tar, to preserve the wood.

TROUGHING can be made with either hollow trees squared down on one side, and laid end to end, or with flat iron tacked into logs hollowed out the required depth to receive the end of each sheet. A rail is laid along each side, and the iron nailed to them. This troughing requires to be protected from cattle.

CHAPTER XX.

RING-BARKING.

UPON this subject the greatest diversity of opinion exists. Scientific men even differ as to the method to be employed—whether the sap should be cut through, or merely the bark; and we daily come in contact with men who advocate one or other of these methods. As this subject is likely to become of considerable importance in this country, too much consideration cannot be expended upon it, nor too great pains taken to ascertain the proper and most advantageous method to be employed before undertaking any considerable quantity of it. It would be most unsatisfactory for anyone, after ring-barking all, or nearly all, his timbered country, to find his labour to have been in vain, by his trees shooting out in young suckers at the place where they have been ring-barked. When this has resulted, the country is even more useless than before, because the object of ring-barking has been defeated, as the trees still require and obtain as much nourishment from the earth as they formerly did; moreover, the country will have been rendered more scrubby, and the scrub will be more difficult to eradicate than the trees would at first have been.

We propose to give the result of our own experience upon this subject, and, as we have devoted a great deal of time and attention to it, we consider it is worthy of some consideration.

In the first place, the time of year for ring-barking depends somewhat upon the locality and climate, and upon the description of timber to be ring-barked. In ring-barking any timber of the *eucalyptus* species, *only the bark*, to a width of *not less* than a foot, should be taken off. The tree, when ring-barked in this manner, will naturally take a longer period to die than if the sap be cut. The sap ascends through the cells and vessels of the outer or sap-wood of the tree, proceeding to the surface of the leaves and utmost extremities of the system, and, having been exposed (chiefly in the leaves) to the influence of air and light, returns through the bark, a portion ultimately reaching the root, and being secreted there, whilst another portion probably enters again into circulation with the new fluid entering from the soil. The elaborated sap always contains much less water than the ascending sap. Much of the sap which is taken up by the roots is, thrown off in perspiration by the bark and leaves. The principal objection to the method of simply barking the trees is the length of time they take to die. We admit they do take longer than if they were ring-barked to the sap, but we maintain it is a more effectual way of *killing* the tree; as, if ring-barked in the proper time of year, *i.e.*, when the sap is up, they will never be the cause of further trouble or annoyance; and they cease to take nourishment from the soil long before they are apparently dead. They sometimes take as long as eighteen months, but we have seen trees quite dead and leafless in a period of ten months after ring-barking. This, however, cannot always be relied upon.

If timber is ring-barked by cutting through the sap-wood, a different appearance is perceptible, and the leaves assume a yellowish tinge in as short a period as twenty-four hours, or almost as soon as if the tree had been cut down. It is owing to this early dying off that many people erroneously think it the better method of the two; but they will invariably have cause to change their opinion before many months have elapsed, and when the green shoots begin to make their appearance below where the axe has been used. Knowing this as we do, it is a matter of surprise to us that this system of ring-barking has any advocates, for we have seen many miles of timber ring-barked in both ways, but we have never seen any that had been ring-barked by cutting through the sap-wood which could at all compare with that ring-barked by taking off the bark only. Trees not of the eucalyptus species may be ring-barked at any time of the year, and in most other trees the method of ring-barking is immaterial. The trees that will be found to die most quickly after being operated upon are box, peppermint, and " yellow-jacket." Of course, every tree that is not sound shows the effects of ring-barking much more quickly than does one that is perfectly sound. In good pastoral country, ring-barking improves the carrying capabilities to the extent of nearly doubling the number of stock.

Some people go so far as to say even quadrupling, but this we have never ourselves seen authenticated. Clumps of trees should be left for shade ; young ones are preferable.   Currajong, sheoak, or any other timber which sheep will eat—and these they eat with avidity— should not be destroyed.   The country should not be heavily stocked for at least twelve months after ring- barking.   If, at any future date, it is deemed advisable to grub and clear any portion of a run, it is more cheaply and easily effected when the timber is thoroughly dead than if it were green, as the surface of the earth need only be removed so as to allow of a fire catching the dry roots, which will generally burn to a sufficient depth below the surface to admit of ploughing.   We have seen many acres of ordinarily timbered country cleared in this way at from 28s. to 30s. per acre.

CHAPTER XXI.

DESTRUCTION OF DINGOES.

THE dingo is considered by some naturalists to be a distinct species of the genus *Canis*, but by others it is regarded as a mere variety of the *Canis familiaris*. It is to be found in all parts of Australia, but there are no distinct traces of its having originated from dogs brought to the colony by man.

The dingo is of a tawny colour; has a larger head than the domestic dog ; the ears are short and erect, and are pointed forward; the tail is brushy; and, in general, he carries his head high and ears erect when running.

In its native state the dingo never barks; but its long, melancholy howl can be heard for miles through the midnight stillness of the Australian bush.

Singly, the dingo is very cowardly; but when collected in large numbers they have been known to attack a man, and it is a common occurrence for them to attack and destroy young cattle. Since the introduction of domestic dogs into the bush, the appearance and character of the dingo have greatly altered. In regard to the former, almost any colour can now be found; and a greater variety of sizes exists. In character it has become much more destructive and daring, and taking, if possible, a greater delight in killing for sport, before making a meal.

As the destruction of this pest is to the squatter a matter of vital importance, many and various means have been tried for its extermination, a few of which we will specify, for the benefit of those who may require the information.

When poison is used in baits, strict attention must be paid to its careful use, as many accidents have happened through carelessness or neglect. The flesh of the emu is looked upon by the dingo as a great delicacy, probably on account of the difficulty they experience in obtaining it. Sardines are also considered a rare tit-bit. In the absence of these, mutton, beef, or opossums may be used— the latter, if possible; but tough or old meat should be avoided. When baits are prepared, if cut small the will be found more effectual. Avoid handling as much as possible; and after inserting the poison (about as much as can be lifted on the point of the large blade of a pampas-knife), if rolled in flour it will be more readily devoured. Many devices are tried to succeed in poisoning dogs too knowing to take a bait laid in the ordinary manner. We have tried the following with success :— Take a tender piece of emu-flesh (or other), and, after inserting the poison, toast gently before the fire, on a fork, care being taken not to touch with the hands. Then take the bait and place it in a piece of paper, to prevent the hands touching it. Take your own dog to the place where the dingo is known to frequent, and with his paws scratch out a shallow hole, and allow the bait to drop from the paper; then, again with your own dog's paws, refill the hole; and if the dingo is to be had with poison, this *ruse* will succeed. Another method of poisoning, which has many advocates, but which our experience does not lead us to put much faith in, is to procure a quantity of rendered suet and place it in a tin (a common lollie tin will do), and mix into it a sufficient proportion of strychnine, before it has had time to cool ; place the tin where the dog is known to run. Where baits are found to be ineffectual (of course in the summer time), poisoned water should be tried. Fill a small vessel with water; then dissolve, in a small quantity of vinegar, strychnine in the proportion of 1 drachm to the pint of water, and lay the poison in the haunts of the dogs, care being taken to prevent stock having access to it.

We would strongly recommend the use of strychnine in preference to arsenic, principally because of the difficulty experienced in the judgment of a proper dose in the administration of the latter; for, if an overdose be given, it will cause the animal to vomit, and defeat its own object; and if too small a dose be given, the result will be the same, in so far as it will not cause the animal's death; and in both cases it causes the recipient to regard baits with suspicion and distrust for the remainder of his lifetime. In the use of strychnine, however, a very small dose will be found effectual, and the greater the dose the quicker the effect.

Besides the various methods of poisoning, many other efforts have been made, having for their object the destruction of this pest—notably, traps. We remember having seen in the *Australasian*, some time ago, a suggestion from a correspondent, which was as follows :—

" Choose a quiet and secluded spot some distance from any public road; build a circle of stakes with a diameter of from 2 to 3 feet, and sufficiently high to prevent a dog from jumping out; then place a second circle of stakes round the other, with a diameter of about 2 feet more than the inner circle ; leave a space in the outer circle of 1 foot in width, and to this have a door arranged to swing in towards the inner circle ; inside the inner circle place a strong lamb, with good bleating proclivities. The dingo, attracted by the bleating of the lamb, enters the outer circle, and, being unable to turn, must go round, and in completing the circle he closes the door. No harm will result to the lamb beyond a little bodily fear."

We have found a very effective remedy in the following plan, which has for its recommendation simplicity:—Take a young dog, one that will howl well, to where it is desired to poison, and fasten him securely by a chain. Then place baits and poisoned water just beyond his reach. His howling will attract the wild dogs, and the chances are that, before leaving, they will take either a bait or a drink. A bitch on heat will answer better in this case than a young dog; of course she must be a worthless one, on account of her liability to being lined by the mongrel dingo.

Another method is to dig a hole or trench along the path which it is known the dingoes traverse, and cover lightly with bark, supported on thin sticks, and sprinkled over with earth. The trench must be dug sufficiently deep to prevent the dingo making his escape when once trapped. We have not much faith in this method, but it may be found of use where the dingoes are not so cunning as it has been our experience to find them. There are dogs much too cunning to allow themselves to be deprived of life by any of the above means, and the rifle has been found to be the only sure exterminator. We have sometimes waited for days, and even weeks, to get a shot; and many times spent the greater part of the night for the dog to come to water.

To dissolve Strychnine.—Besides the plan we have given for dissolving strychnine, the editor of the *Australasian*, in answer to a correspondent, gives the following:—

"To dissolve 1 oz. of strychnine add, say, 3 ozs. of muriatic acid and about 2 pints of water; mix, and dissolve by heat, and while hot add sufficient hot water to make one gallon. Each 8 ozs. of solution will contain 1 drachm of strychnine, 2 grains of which—or its equivalent, 2 drachms of solution—is sufficient to poison any dog."

CHAPTER XXII.

MARSUPIALS AND THEIR DESTRUCTION.

PERHAPS there is no native animal in the colonies whose presence and habits have put the squatter to more annoyance and expense than the kangaroo. Indeed, in most parts of the colonies this may be stated more as an established fact than as an hypothesis; but away in the back country the plague of dogs is doubtless a source of more trouble to the pioneer than the presence of the marsupial denizens of the forest.

Sheep dislike feeding after kangaroos as much as cattle do after sheep, and for this reason they are very undesirable, to say nothing of the grass and water they consume, and the interminable injuries they inflict upon wire fences. If no means are taken for their destruction, their increase seems to progress in an equal degree with that of the stock; or, rather, as the country is improved and watered, and the grass thickened and strengthened by the introduction of stock, the kangaroos increase rapidly in numbers. This is easily accounted for; as, for one thing, the fences erected prove a considerable barrier to their migration, and the presence of plenteous supplies of grass and water removes their desire or necessity for change of locality.

The destruction of kangaroos has been for some years compulsory in Queensland, and by a late Act of the New South Wales Houses a similar law is now in force in that colony.

By this Act all stockowners have to pay an assessment, *per capita*, for all sheep and cattle depastured on their runs; and in Queensland this assessment is varied, and depends upon the estimated number of the marsupials in the district in which the station to be taxed is situated. In New South Wales the rate is uniform, and this fact has caused considerable discontent in those portions of the colony but thinly populated by kangaroos.

The sum thus collected is used as a fund to defray the expenses of extermination. Each district is presided over by a board and an inspector, and receivers of scalps are appointed all over the colony. These receivers count the scalps brought to them, and forward a certificate to the inspector, who, at the monthly meeting of the board, obtains cheques for the various amounts due; the rate in New South Wales being sixpence per scalp, and in Queensland from twopence to ninepence, according to the different species of marsupials destroyed. It is the duty of the receivers to destroy the scalps, after having ascertained their number.

According to the Act now in force in New South Wales, the Government may, in the event of any squatter neglecting to destroy the kangaroos, send men to effect the necessary destruction, and charge the cost of such work, with all its consequent expenses, to the lessee of the run.

Many plans for the destruction of this pest have been tried. On the Darling Downs, and in other places where kangaroos are numerous, many methods have been made use of for their extirpation—amongst others, yarding. To effect this, considerable expense must be gone to; and the plan, although effective, will be found too expensive, unless the marsupials exist in large numbers in the vicinity. Strong yards are built, very high, and with long wings to them. They should be erected on a line of fence, and in many cases may be so placed as to afford means for the destruction of the marsupial inhabitants of four paddocks, and always of two. The yards must be of good stout timber, and no little economies in their erection should be practised; let them be most substantial, and equal to a considerable pressure. The wings may be of long strips of white calico stretched from tree to tree, or, in the absence of these, from post to post. They are not intended to resist the weight of the kangaroos, but merely to frighten them into the course it is desired that they should go. When the yards have been made ready, a number of men on horseback should enter the paddock and proceed to work in almost a similar manner to a sheep muster, with plenty of noise. As the yard is approached, the kangaroos must be quickly followed up, to prevent the forerunners from turning back after having scented danger; and as many

as can be are now forced into the yards. Their destruction may now be effected either by shooting, or, less
expensively, by clubbing.

Shooting is also a common method of extirpation, and
seems daily to come more into vogue, many practised
shots making fair wages by this means.

Poison has also been tried as a means of destruction,
and is said by many to be very effectual if pursued in the
following manner :—

The first requirement for this system is the complete
absence of stock from the paddock to be operated upon.
A few pieces of rock salt are scattered about a hill, as
though for the use and benefit of sheep. Some coarse
Liverpool salt is then poisoned by mixing with it powdered strychnine, in the proportions of 1 drachm of
poison to 1 pound of salt, and placed in a receptacle
that will not allow it to escape on to the ground, as that
would endanger the lives of stock afterwards turned into
the paddock; and, as kangaroos are very partial to salt,
they will partake freely of the mixture, and soon die.
Too much rock salt must not be placed about, as all that
is required is sufficient to entice the marsupials.

As kangaroos do not care to partake of water when in
small quantities, poisoned water, such as would be used for
native dogs, will not be found so successful.

There is some demand for the skins of kangaroos, both
in the colonial and English markets; and, where the
charges for carriage are not excessive, this branch of commerce has been found tolerably lucrative. The hides,
when tanned, are principally used for the manufacture

of gentlemen's riding boots, but are also used for many other purposes.

Outside of Australia a halo of romance encircles the kangaroo. In many parts of the world he is spoken of with dread, and accredited with ferocity almost equal to the tiger; with strength but little inferior to that of the elephant, and with a bloodthirsty disposition, which can only be compared to that of a lion. His rage (once roused) is believed to be terrible; and altogether he is supposed to be a most formidable animal, and quite capable of making a meal off two or three human beings.

Many hair-breadth escapes from him are to this day to be found recorded in the school-books of both England and America, some of them extending over a period of several hours. This is, of course, as all Australians know, a most incorrect and exaggerated view of the matter. The kangaroo is naturally of a timid and retiring disposition, and prefers running away to fighting, when he is allowed his choice.

When actually brought to bay, many of them do show fight, and some "old men" are by no means mean antagonists; but in most cases a man with a stout stick is more than a match for them.

We must also remark that the kangaroo does not use his tail as a means of progression, by passing it under himself and springing from it, as is believed by some.

CHAPTER XXIII.

WILD HORSES, OR " BRUMBIES."

UNDER this heading we have but very little to say. Had we been writing a few years ago, we would, perhaps, have had much more to write about. In some parts of New South Wales, many squatters have gone to a good deal of expense to rid themselves of this nuisance. We know of one large station in Riverina where upwards of 2,000 head were destroyed, at a cost to the squatter of 3s. per head. Besides this price, those engaged in destroying, which was done in this case with the rifle, were allowed the hide and hair—in this part of the country worth about 6s. By this it will be seen that very fair wages could be earned by the use of the rifle.

Other methods besides the rifle have been adopted for the destruction of wild horses, such as yarding. This is done by building strong yards, and fencing the water off, until the brumbies become weak and exhausted for the want of water; they are then run into the yards without much trouble. Any good young horses that may be among them are selected, branded, and broken in, and the remainder destroyed or impounded. Where practicable, this method is, of course, more profitable than shooting the horses indiscriminately would be. We have often seen some very excellent horses run in from the bush in this manner; and when taken young they are no more difficult to break in and handle than are ordinary station-bred horses. We have seen horses running in the bush that could have been had for the trouble of yarding; but when this has been attained they have brought prices ranging as high as £15 to £25 per head. These, however, are generally horses that have been well bred, and escaped unbranded.

CHAPTER XXIV.

STATION EMPLOYES.

AMONG many squatters and managers it is a matter of surprise that men cannot be found who will remain in their employ any length of time. The men are generally blamed, and sometimes justly so; but if we, who are comfortably housed, will take a glance at the filthy hovels and "pigstyes" many employers provide for their men, we will find that these badly ventilated and miserable abodes have much to do with the discontent of the men, and probably cause much of the erratic disposition we hear so much about. Apart from the miserable accommodation provided in many cases for station hands, the latter have, in our opinion, good grounds for complaint of the general treatment they receive at the hands of their employers, some of whom treat their men as though they were slaves, seldom speak to them a kind word, and almost invariably address them in terms which a gentleman would not use to his dog. If such employers as these could procure men totally devoid of self-respect, and dead to all feeling, they might find it less difficult to retain, more permanently, their servants.

No right-thinking man could, however, wish to see this state of affairs inaugurated; and it is only just to observe that the minority of managers treat their men in the cruel and unjust manner we have alluded to, and this minority consists principally of men who have risen from poverty to positions of comparative wealth.

Est modus in rebus; and although a certain amount of firmness and fixedness of purpose is necessary to sustain the dignity of authority which every employer should possess, yet this should be tempered with a degree of kindness and civility quite consistent with the prestige of rule, which will create a bond of union between employer and employed. It has ever been our pleasant experience to find that where such good feeling as this existed, the work was always done in a better spirit, and in a more satisfactory manner, than where the employer, in constant fear of being deceived, considered it necessary to watch and nag at his men from morning till night.

It cannot be denied, however, that there are men who would impose on their employers if they could. It is, of course, the duty of every employer to see that the work is properly done; yet this is sometimes impossible without a system of espionage, which is repulsive to any but a coarse-minded man, and one, in fact, who would require the same surveillance himself. By the experienced employer such men can be generally distinguished at first sight; but it sometimes happens that even the most experienced finds himself deceived by appearances, and discovers that instead of the honest, intelligent, hard-working man whom he thought he had engaged, he is giving employment, or at least wages, to an undeserving scoundrel, who spends the greater part of his time "lounging" or "sundowning"—*i.e.*, spending his time under a shady currajong, or other sleep-inspiring tree. We need scarcely add that such men should be dismissed immediately, as, besides being thoroughly useless themselves, they spread their indolent habits among their fellow-servants.

To the inexperienced manager this servile species of mankind is a snare and a deception, as he works like a Trojan to his face, and as soon as his back is turned he is the first to rest from his Herculean efforts. His life is one continual living lie, and he awaits every opportunity to deceive his employer, to whom he crawls and cringes as a mongrel dog to his master. There is a class of men, in the positions of managers, whose long experience does not secure them from deceptions of this kind, who form hasty and incorrect estimates of the characters of their men, and require more than sufficient proof of their incapacity and indolence. This same hasty judgment often-times leads to the discharge of good, hard-working men, and the retention of those who are inefficient. As one of the principal particulars in the management of a station consists in having a good staff of reliable men— men who may be trusted, and who know they are trusted; for in this last essential, to a knowledge of the confidence placed in them is to be traced much of the good feeling which should ever exist between master and man—every effort should be made to retain them.

CHAPTER XXV.

HUTS AND OUT-BUILDINGS.

ANY hut or out-building should be built on rising ground, the side of a hill being a suitable site generally. They should be at convenient distances from the homestead. The men's hut should be of a roomy and substantial character. The berths, or "bunks," should not be placed more than two high; the too common method of placing them one on top of the other four in height is not to be commended, as it is productive of ill health, and is by no means conducive to comfort. We have already said huts should be of a roomy structure—they should also be well ventilated, and should be provided with a commodious fireplace, and, in short, with all such possible comforts as will enable the men to spend their Sundays and evenings in as home-like a manner as possible, and so give them every inducement to spend their leisure hours on the station, instead of galloping off to the nearest public-house or shanty to look for the comforts otherwise denied them. The huts should be swept out every morning, and, towards evening, a fire lighted. This is generally done by the man employed for milking, wood-cutting, running up horses, &c., &c., generally to be found on most well-managed stations.

There are various ways of constructing huts and out-buildings, and the matter of shape is, of course, entirely a matter of taste; into this we need not enter. Huts may be built of what is termed round horizontals—the posts are first bored with a 2-inch auger, and a groove formed the whole length of the post above ground ; this may be done, but not so securely, by nailing two 1½ by 2 inch battens, 2 inches apart, on the sides of the posts, squared to receive them, and which are placed at a depth of 2 feet in the ground, at convenient distances—say from 4 feet to 6 feet, and the round timber dropped horizontally into the grooves. A fire-place can be built in the same manner, and stoned inside; or, where stones are not procurable, clods of turf may be used, and will be found to last very well if continually white-washed. The rafters and battens may also be of round timber, flattened on one side, and the roof and gable ends of bark, which can be fastened to the battens with hide quite as securely as, and less expensively than, with nails. This constitutes a good substantial bush hut, but in the more settled districts sawn timber or bricks should be used, with corrugated iron for roofing.

A good hut may be built out of the refuse half-slabs from a saw-pit, by placing them vertically instead of horizontally, one row being placed with the round side in and close together, and other slabs laid over the crevices thus formed, with the round side out. This is even a less expensive structure than the horizontally built hut.

CHAPTER XXVI.

FELLMONGERING.

FELLMONGERING is, upon stations, generally carried on in a very primitive manner; indeed, the very limited amount of this work that is done on the station renders it necessary that it should be carried through in the most expeditious and least costly manner possible. We have already observed that, if done in this way, it becomes more profitable to fellmonger the skins than to send them to market in their natural state, especially if the distance to market is great.

Fellmongering should be done at or as soon after shearing time as possible, so as to allow of the fellmongered wool being sent to market with the general clip.

The skins should be soaked in water for about twelve hours, and then stacked in heaps—fleshy sides and woolly sides together—and left so until the wool can be easily taken off by plucking. The wool is then washed in cradles made for the purpose, and dried on calico or tarpaulins. If soaked in warm water, in which has been dissolved a small quantity of soft soap, prior to washing, it will add considerably to the cleanliness of the wool.

Wool that has been washed should not be pressed until thoroughly dry.

CHAPTER XXVII.

SHEEPSKINS AND HIDES.

THE reason that station skins command such a low price in the market is because they are thought so little of, and are seldom taken any care of on stations. We have seen many stations where the skins have been allowed to rot, and thus become almost useless; this is frequently the case, but there are even places where the hides of the animals slaughtered for beef are buried with the offal. Where such waste and extravagance as this exist, prosperity is never deserved, and need scarcely be looked for.

Sheep should never be killed for mutton with a growth of wool of more than six months—that is, on stations where cattle are obtainable,—as after that period it becomes unprofitable, owing to the small price to be had for station skins as compared with the value of wool sold with them. As soon as the sheep is skinned, to preserve the skin it should be hung lengthwise over a rail, and not exposed, under any circumstances, to the rays of the sun, or to wet weather. The skins should not be laid one on top of another until thoroughly dried, and even then they should not be laid more than two deep, and require to be examined occasionally. If skins were treated in this manner, much better prices could be demanded for them in either the metropolitan or local markets. A solution of arsenic and water, laid on with a brush on the fleshy side, will preserve the skins from weasels, &c.

Where suitable accommodation and appliances are obtainable, we would advocate the system of fellmongering, especially to those who are a distance from market. But, for the benefit of those who prefer the more common practice of sending their skins to market in their natural state, we recommend the following plan of securing the skins in compact and neat bundles, saving the expense of wool bales :—

On the floor of the wool-press, and running under-

neath the side doors, lay two pieces of ordinary fencing
wire, at equal distances from the sides of the press.
Upon these lay three pieces of 6 by 1 inch boards, the
length of the width of the press.   Fold the skins woolly
sides in ; double, and pack the press full, they being well
trodden. When this is done, place three boards upon
the top of the skins similar to those on the bottom, and
upon these two pieces of wire the same length as those
on the bottom. The monkey is then run down, and
when the skins are pressed the side doors are thrown
open, the top and bottom wires securely fastened together,
the monkey raised, and the skins are not only in compact
bundles, but are more securely and more economically
bound than if pressed in the ordinary wool bales. The
hides of cattle, instead of being buried with the offal,
will be found more profitable if spread out upon the
floor in some spare room or outhouse, and sprinkled with
coarse Liverpool salt well rubbed in with the feet or
hands ; they should be laid one on top of the other,
hairy side and fleshy side alternately together. When
prepared for transport to market they are spread out
singly, the two flanks turned in to meet in the centre,
doubled again, and rolled tightly from the head to the tail,
which makes a strong and secure tie for each separate skin.

CHAPTER XXVIII.

SHEEP-DOGS.

THE sheep-dog is the most useful and valuable of all

kinds of dogs, and universally used throughout Europe, and in the countries colonized from Europe, and also in some parts of Asia, to assist them in the tending of their flocks. Without it the shepherd would be utterly incapable of taking care of the great number of sheep often under his charge; and the expense of keeping the requisite number of shepherds would far more than take away the profits of sheep-farming. That the dog was employed in the tending of sheep in very ancient times we learn from the allusion to *the dogs of the flock*, in Job xxx. 1. Buffon imagined the shepherd's dog to be the original of all the domesticated dogs, but was unable to assign any good reason for such an opinion. The shepherd's dog exhibits nearly the same characters in all parts of Europe, although there are slight diversities in different countries—as between that of England, and that of Scotland, there known as the *collie*. It is of middling size—differences of size, however, being amongst the characteristics of different races; of rather slender form, with a pretty sharp muzzle; the ears erect, or, in some cases, drooping at the tip; the hair soft, long, shaggy, and somewhat waved; the tail slightly pendulous, more or less recurved, and very bushy; the feet well protected by hair, so as to be adapted for rough ground. The eye is very bright and intelligent, although the ordinary demeanour of the animal is remarkably calm and quiet. No kind of dog is more intelligent, and perhaps none so docile. Its ready comprehension of the meaning of its master, its prompt obedience to his word or gesture, its evident knowledge of what is requisite to be done, and the services which it performs, can never be observed without admiration. A shepherd's dog exhibits the utmost care to prevent sheep from straying off the road along which they are being driven; and sets itself, often of its own accord, to watch any gate or gap in the fence, or goes immediately to bring back stragglers. It is equally useful on the bleak moor or wild mountain—readily going for sheep, and bringing them from a distance. The sheep become perfectly acquainted with it, and evidently regard it as a friend, and not as an enemy; although the appearance of any other dog would alarm them at once. It knows the sheep of the flock it is required to attend, and, even in a crowded market, adroitly separates them from others with which they have become mingled. Its remembrance of places is obviously accurate; and a dog which has found great difficulty in conducting sheep through crowded thoroughfares does the same work much better on subsequent occasions. The intelligence of the shepherd's dog has sometimes been proved in a remarkable way by dishonest masters employing them to steal sheep—the master merely indicating, by some sign, the sheep which he wished to add to his own flock, and leaving the dog to do it in his absence. For stealing in this way, a farmer in the south of Scotland was hanged about the end of last century. More frequent instances are on record of the shepherd's dog conducting a flock of sheep safely home, for many miles, unaccompanied by the shepherd. The shepherd's dog is affectionate, and becomes strongly attached to its master, but is generally shy to strangers. It is generally treated with great gentleness by the shepherd; no severity is used in its training, nor could be used with advantage. It is very muscular and active, and capable, perhaps beyond any other kind of dog, of continuing its exertions for a long time.

The shepherd's dog is often crossed with other kinds of dog; and dogs thus obtained which are capable of all the services required by shepherds.

CHAPTER XXIX.

PINE AND OTHER SCRUB.

WE have not sufficient space at command to enter, in this place, as fully into the consideration of this subject as we could wish. Were it not for the interest attached to it through certain recent action on the part of the authorities in one of the colonies, we would not enter into it all—the limits of the subject being somewhat outside the compass of this work. Under the circumstances, however, and having cognizance of the great interest now attached to the growth of pine scrub in particular, and the certain future prominence of the matter, we deem it wise to devote some attention to it, be it but brief.

An Act of the Parliament of New South Wales, passed some years ago, prohibited the ring-barking or cutting down of pine and other trees, under certain sizes; and only allowed the prosecution of such work under approval and permission. Recent action on the part of the Minister has, however, deprived the leaseholder of even this last favourable clause; and the ring-barking of trees and cutting of scrub has been entirely prohibited, under heavy penalties. To the consideration of the wisdom of this action we will here confine our attention.

For many years the squatting community of a great part of Riverina have viewed with alarm the rapid and thick growth of young pines. In the back blocks of the Lachlan this has been especially the case; and many places—portions of the country which, from three to six years ago, presented the appearance of well-grassed plains, or open forest country—are now covered with a dense growth of young pines, through which it is difficult to ride, and among which no grass can be found. In many other parts the growth of young pines, though not sufficient to render the country utterly useless for pastoral tenure, is still more than adequate to cause alarm.

The cause of this growth has been made matter for considerable speculation, and various theories have been advanced regarding it. Perhaps the most strikingly original of these is that put forward by the writer of an article in the *Town and Country Journal* of 16th July, 1881. The writer sets forward the view that the growth of young pines is attributable to the practice of ring-barking; that the seeds from the trees are induced to fall, from deprivation of life; and that sheep depastured on the land receive myriads of these seeds in their wool, which they drop in their path while feeding. The writer also states that in parts of New South Wales dense forests of pine scrub now grow where, some years ago, the land was fine, well-grassed plain country, with here and there belts of pine trees.

This view of the matter is, we confess, quite new to us, and we cannot, in the face of what we know to be facts, place any reliance upon it. We will briefly re-capitulate those facts.

In the first, the seeds of the pine tree attain their mature and ripe state while still adhering to the trunk, or, rather, branches of the tree; but, in that stage of their existence they are very easily removed, core and all; and at the foot of any pine tree, over the age of five years, these cones, containing the seeds, and opening to allow of their exit, may be found in large numbers. Apart from this fact, however, supposing the case of a stretch of plain country with here and there belts of pine, no manager would for a moment think of ring-barking these trees; the shade they afford would be much too precious to allow of it. The timber might be required for building or fencing purposes, but under no other circumstances would the trees be likely to be meddled with. There are also innumerable localities, such as we have before cited as plains from three to six years ago, now rendered utterly useless from the inordinate growth of pine scrub, and in these localities the timber has never been ring-barked. There are, again, places where neither ring-barking nor felling of trees has been proceeded with, and where no stock—either

sheep or cattle—have been depastured, and these present
a growth of young pines sufficiently thick to render the
country valueless for pastoral pursuits in the course of a
very few years, if something be not done to destroy them.
Finally, we can point out country that has been ring-
barked, and has also carried stock, and that to a greater
extent than its natural capacity would have allowed of,
where the young pines are mainly conspicuous by
their absence. It is impossible to reconcile these facts
with the theory we have referred to; and, whatever the
causes of the inordinate growth of scrub in parts of the
colonies may be, we feel satisfied that it does not in any
degree arise from the practice of ring-barking. On the
contrary, we are certain, and so must all those be who
have any practical knowledge on the subject, that ring-
barking forest land is most beneficial to that land, and is
a practice which should be protected and encouraged,
within certain wise limitations, by our legislative
assemblies.

The growth of young pines in the vicinity of older
trees of the same species is natural; it is what we must
expect. We do not, however, believe the growth to be
much more extensive now than it was twenty years ago.
In our opinion, the seeds, after ripening, are disengaged
from the cone, and that from the tree bearing it, by the
wind; and the same agency is, we believe, responsible
for the growth of young pines at various, and sometimes
great, distances from the parent root. If this be
admitted, it will be asked, why did not this young
growth of pines attract the attention it now does, many

years ago? And this we may justly account for by the fact that, years ago, large and fierce bush fires raged periodically, destroying all such young growth; and that now fires are very rarely known, and, when known, are extinguished with what speed is possible.

That the growth under discussion should be fostered and encouraged by our Government says but little for the intellects of those who rule us. Even if preserved for future building purposes, these young trees require thinning; in their natural state they can come to no useful maturity, the stems being too crowded to admit of healthy growth.

We are by no means in favour of indiscriminate ring-barking. Far from it; there is a medium in all things, and it cannot be denied that, in many parts of the colonies, great and irremediable damage has been done through the reckless and hasty action of some lease-holders.

We have not space here to enter into the side view of the question, regarding the probable effect, from such action, upon the rainfall of forest country. This is admittedly a moot question; but, apart from this, the total destruction of our forests would indeed be most unwise. Timber is required for building, for railways, and for many other purposes, and certain portions of our forests should be reserved for such requirements. Let these portions be judiciously selected, that they may in future be sufficient for our wants. Certain limits as to distance might be arranged for the preservation of timber about townships, or probable sites for future towns.

Throughout the colonies, areas might be chosen and preserved by Government from the ravages of the axe; while the remainder of the country should and would be cleared of its present unhealthy, useless undergrowth, and be thus preserved for grazing purposes.

Walker, May and Co., Printers, 9 Mackillop-street, Melbourne.

INDEX.

www.ingramcontent.com/pod-product-compliance
Lightning Source LLC
Chambersburg PA
CBHW081716220526
45468CB00008B/1869